Feedback Linearization of Dynamical Systems with Modulated States for Harnessing Water Wave Power

Synthesis Lectures on Ocean Systems Engineering

Editor
Nikolaos I. Xiros, *University of New Orleans*

The Ocean Systems Engineering Series publishes state-of-the-art research and applications oriented short books in the related and interdependent areas of design, construction, maintenance and operation of marine vessels and structures as well as ocean and oceanic engineering. The series contains monographs and textbooks focusing on all different theoretical and applied aspects of naval architecture, marine engineering, ship building and shipping as well as sub-fields of ocean engineering and oceanographic instrumentation research.

Feedback Linearization of Dynamical Systems with Modulated States for Harnessing Water Wave Power
Nikolaos I. Xiros
2020

Marine Environmental Characterization
C. Reid Nichols and Kaustubha Raghukumar
2020

Feedback Linearization of Dynamical Systems with Modulated States for Harnessing Water Wave Power
Nikolaos I. Xiros

ISBN: 978-3-031-01363-8 paperback
ISBN: 978-3-031-02491-7 ebook
ISBN: 978-3-031-00319-6 hardcover

DOI 10.1007/978-3-031-02491-7

A Publication in the Springer series
SYNTHESIS LECTURES ON OCEAN SYSTEMS ENGINEERING

Lecture #1
Series Editor: Nikolaos I. Xiros, *University of New Orleans*
Series ISSN
Synthesis Lectures on Ocean Systems Engineering
ISSN pending.

Feedback Linearization of Dynamical Systems with Modulated States for Harnessing Water Wave Power

Nikolaos I. Xiros
University of New Orleans

SYNTHESIS LECTURES ON OCEAN SYSTEMS ENGINEERING #1

ABSTRACT

As pointed out by other researchers, hybrid structures in ocean engineering are based on flat concrete foundations. Due to wave action these foundations are exposed to different pressure distributions on the top and bottom sides. As a result, the bottom side is exposed to a saddle type pressure distribution leading to huge forces on the foundation. Indeed, such huge forces have been observed at a number of offshore platforms installed in the North Sea.

In an attempt to turn a problem into an advantage, the concept in this work aims to develop an integrated system to harness and harvest ocean wave energy right at the seabed. The long-term interest is to develop integrated devices that can be used as actuators or sensors, which, due to low manufacturing cost, can be employed in large quantities for control of ocean engineering systems, e.g., maritime renewable power-plants, or monitoring of marine processes, e.g., oceanographic sensing.

A key element to the proposed system is the nonlinear coupled electromechanical oscillator unit, the dynamics of which are investigated with a novel approach in this work. The fundamental nature of the oscillator at hand makes it an excellent choice for applications involving oceanic transducers consisting of a dry driving electrical stator physically separated from a wet-driven payload mechanism. Without such units available at a low cost and a large number, harvesting the energy of a vibrating plate at seabed may prove impractical.

KEYWORDS

exact, feedback, linearization state-space, modulation, demodulation

Contents

List of Figures

Acknowledgments

The author would like to express his gratitude to all the mentors that have been nourishing and advancing his intellectual and academic achievements through his life and career. Without their invaluable guidance, advice, and faith in him, this work would not have been possible.

Also, the author would like to thank his family, and especially his wife, Fay Kalergi, for the day-to-day practical support in the myriad of things that arise in life. Also, the author has deep appreciation for their companionship and patience while conducting this work.

Finally, to all those people that the author may forget to acknowledge but have contributed in the successful completion of the work in hand.

Nikolaos I. Xiros
March 2020

CHAPTER 1

State-Space Modulation and Demodulation

1.1 MOTIVATION AND BACKGROUND

The analysis of coupled dynamic systems, including nonlinearly coupled but spectrally separated and effectively linearly decoupled (as demonstrated later on) subsystems, has gained significant attention recently. This is due, at least partly, to the developments and associated needs in the disciplines of telecommunications and mechatronics, e.g., [1–6]. In telecommunications engineering, carrier, e.g., amplitude, modulation is required for the transmission of digital or analog information over a channel which may be a conductor, or a waveguide, or in the case of wireless communications, free space. Furthermore, utilization of the channel is rather commonly required to be separated into a number of mutually exclusive user pairs which desire to communicate independently and without interference due to either external noise sources or crosstalk from other communicating user pairs. In Fig. 1.1, a generic model of a communication channel as a distributed parameter electromagnetic system is shown. The model can be used for either wire-line or wireless transmission and with appropriate adaptations even in the case of the optical fiber medium. The governing principles are Maxwell's equations, rather than the ones applicable to lumped circuits, where Kirchhoff's loop and node algebraic equations suffice. This situation arises in most cases, depending on the frequencies employed and the physical dimensions of the system, where the wavelength of the electromagnetic waves may be relatively small. Otherwise, electromagnetic interference (EMI) due to crosstalk from neighboring elements or lines may occur in a rather broad spectrum, depending on the overall design of the system.

However, the most important and rather common feature of all widely used channels is their frequency selectivity. Indeed, the bands where transmission with relatively low attenuation is possible are limited both in number as well as in extent. Therefore, modulation of the carrier wave by the information signal must be employed. Modulation is practically relocation of the information signal spectral content to a band which is mostly suited for information transmission over a specified channel. The simpler, yet most usual, modulation method is amplitude modulation.

In the wider context of mechatronics, electromechanical systems analysis and synthesis, especially in comparatively small scales, has gained significant interest, due to developments in the rapidly evolving field of MEMS/NEMS (Micro-Electro-Mechanical Systems, Nano-Electro-Mechanical Systems) [1]. The final objective is to develop integrated devices that can

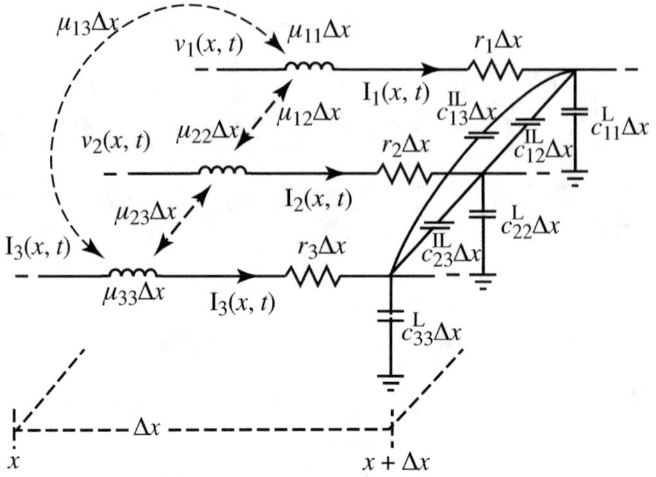

Figure 1.1: Electromagnetic telecommunication channel model.

be used as tiny actuators or sensors, which, due to low manufacturing cost, can be used in large quantities for monitoring and control of complex systems and processes, like, e.g., structural health monitoring of aircraft, spacecraft, watercraft, automobiles, etc., as well as industrial plant monitoring and control. In the typical case, a small-scale mechanical system like a resonator or a spring-mass-damper interconnection is driven by analogue electronics, like voltage or current sources, amplifiers, etc., which, in turn, are accurately controlled by digital embedded components like FPGAs (Field-Programmable Gate Arrays), DSPs (Digital Signal Processors), etc. that actually implement the intelligence in the device. Parameters like silicon chip surface utilization, power consumption, and level of integration are of utmost important for commercial and technological success of a newly proposed design.

Nonetheless, a rather important feature in the analysis of such systems comes from the strong nonlinearity commonly appearing in the coupling between spectrally decoupled parts of the system. As seen in the analysis, even in the case of low frequencies, where lumped models can still be employed the nonlinear coupling between parts of the system requires specific treatment, using advanced mathematical tools [1, 2].

In this context, an alternative, frequency-domain state-space approach is pursued here. In the rest of this work, a specific class of systems with structure comprising linearly decoupled but nonlinearly coupled subsystems is examined. The mathematical toolbox of the Hilbert transform is appropriately introduced for obtaining two low-pass subsystems that form an equivalent description of the essential overall system dynamics. The procedure is then applied to an arrangement commonly encountered in mechatronics and energy harvesting. In this arrangement, a voltage or current source is coupled to a mechanical second-order oscillator, consisted of a mass-damper-spring interconnection, through an electromagnet [1, 3–7]. Such an arrange-

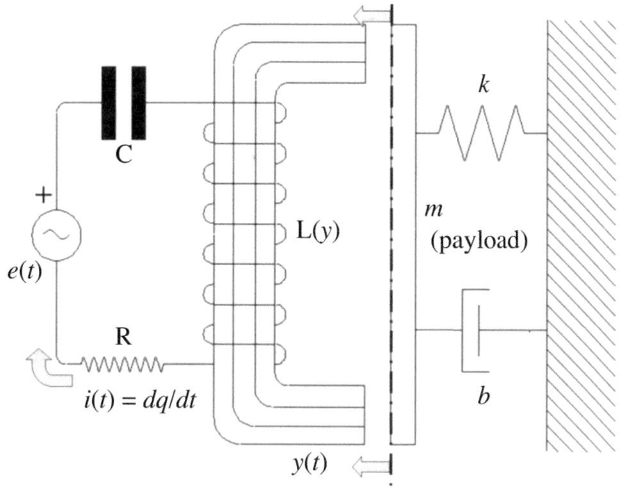

Figure 1.2: Coupled electromechanical oscillator.

ment is shown in Fig. 1.2. A voltage source is driving the RLC circuit which is coupled to the mechanical oscillator through the inductor's armature. In this configuration, the voltage source plays a dual role: it is a source of power that allows the mechanical part of the system to move and, at the same time, of the information control signal that drives the mass (payload) of the mechanical subsystem to the desired position. Furthermore, the electromechanical coupling plays a dual role as well. Indeed, a mutual interaction is established. The movement of the metallic mass induces a change to the value of the inductance apparent to the rest of the driving electric circuit. In this system the low-pass part is the mechanical oscillator and the band-pass part is the electric circuit that drives the electromagnet.

1.2 THE GENERAL SYSTEM FORM IN STATE SPACE

As demonstrated later in a detailed example, a general class of dynamic systems will be considered. The description of this class of systems is cast in a state-space framework. Let vector $\mathbf{x}$ denote the full state vector of the coupled system. Based on a properly defined partitioning of the coupled system's state vector, this description admits the following mathematical structure:

$$\dot{\mathbf{x}}_1 = \mathbf{A}_{LP} \cdot \mathbf{x}_1 + \boldsymbol{\psi}(\mathbf{y}_2) + \mathbf{d} \tag{1.1}$$

$$\begin{aligned} \dot{\mathbf{x}}_2 &= \mathbf{F}(\mathbf{x}_1) \cdot \mathbf{x}_2 + \mathbf{G}(\mathbf{x}_1) \cdot \mathbf{u} \\ \mathbf{y}_2 &= \mathbf{C}_2 \cdot \mathbf{x}_2 . \end{aligned} \tag{1.2}$$

The various terms entering the above equations are defined as follows:

$$\begin{aligned}
&\mathbf{x}_1,\ \mathbf{d} \in \mathbb{R}^{n_1},\ \mathbf{A}_{LP} \in \mathbb{R}^{n_1 x n_1} \\
&\mathbf{x}_2 \in \mathbb{R}^{n_2},\ \mathbf{u} \in \mathbb{R}^{m_2},\ \mathbf{y}_2 \in \mathbb{R}^{p_2},\ \mathbf{C}_2 \in \mathbb{R}^{p_2 \times n_2} \\
&\boldsymbol{\psi} : \mathbf{y}_2 \in \mathbb{R}^{p_2} \to \boldsymbol{\psi}(\mathbf{y}_2) \in \mathbb{R}^{n_2} \\
&\mathbf{F} : \mathbf{x}_1 \in \mathbb{R}^{n_1} \to \mathbf{F}(\mathbf{x}_1) \in \mathbb{R}^{n_2 \times n_2} \\
&\mathbf{G} : \mathbf{x}_1 \in \mathbb{R}^{n_1} \to \mathbf{G}(\mathbf{x}_1) \in \mathbb{R}^{n_2 \times m_2}.
\end{aligned}$$

The partitioning of the n-dimensional state vector $\mathbf{x}$ into two components $\mathbf{x}_1$ (dimension n_1) and $\mathbf{x}_2$ (dimension n_2), respectively, is reflecting the partitioning of the system to a low-pass (LP) and a band-pass (BP) part. This is made clearer by elaborating on the motivation leading to the introduction of LP and BP systems.

To this end, for Single-Input-Single-Output (SISO) linear, asymptotically stable systems, whose dynamics can be defined by a scalar transfer function without poles in the right-half s-plane or on the imaginary axis, the following LP system definition is given:

$$\forall \varepsilon > 0, \quad \exists W(\varepsilon) > 0 : |\omega| > W(\varepsilon) \ \Rightarrow\ |H(\omega)| < \varepsilon. \tag{1.3}$$

Real-valued function $H(\omega) = H(s = j\omega = j2\pi f)$ stands for the transfer function of the SISO system. The parameter $\mathrm{BW} = (2W)$ with W such that $\varepsilon = \frac{\|H(\omega)\|_\infty^2}{2} = \frac{\left[\sup\limits_{\omega \in \mathbb{R}} (H(\omega))\right]^2}{2}$ is called half-power bandwidth or 3 dB bandwidth (or simply bandwidth if there is no chance for confusion) of the system.

Generalization of the LP system leads to BP ones defined as follows:

$$\exists \omega_c, \quad \forall \varepsilon > 0, \exists W(\varepsilon) > 0 : |\omega \pm \omega_c| > W(\varepsilon) \Rightarrow |H(\omega)| < \varepsilon. \tag{1.4}$$

In analogy, a bandwidth BW is defined for BP systems. Angular frequency ω_c is called the carrier frequency of the system. Evidently, LP systems are BP systems with $\omega_c = 0$.

In the case of linear, asymptotically stable, Multi-Input-Multi-Output (MIMO) systems the above definitions may be straightforward generalized, with respect to the transfer function matrix $\mathbf{H}(s)$ of the system and its frequency-dependent maximum Singular Value $\sigma_{\max}\{\mathbf{H}(\omega)\}$. For example, the BP system definition may be extended in the MIMO case as follows:

$$\exists \omega_c, \quad \forall \varepsilon > 0, \quad \exists W(\varepsilon) > 0 : |\omega \pm \omega_c| > W(\varepsilon) \Rightarrow \sigma_{\max}\{\mathbf{H}(\omega)\} < \varepsilon. \tag{1.5}$$

Finally, it is mentioned the definitions of BP and LP systems can straightforwardly be extended to finite energy, scalar or vector signals.

To carry out the analysis of the partitioned formulation of the system, Taylor expansions around zero for the typical multivariable vector and matrix functions will be employed. In particular, we have

$$\begin{matrix} \boldsymbol{\psi} \\ \mathbf{F} \\ \mathbf{G} \end{matrix}(\boldsymbol{\xi}) = \begin{matrix} \boldsymbol{\psi}_0 \\ \mathbf{F}_0 \\ \mathbf{G}_0 \end{matrix} + \begin{matrix} \boldsymbol{\psi}_1 \\ \mathbf{F}_1 \\ \mathbf{G}_1 \end{matrix}(\boldsymbol{\xi}) + \begin{matrix} \boldsymbol{\psi}_2 \\ \mathbf{F}_2 \\ \mathbf{G}_2 \end{matrix}(\boldsymbol{\xi}\otimes\boldsymbol{\xi}) + \begin{matrix} \boldsymbol{\psi}_3 \\ \mathbf{F}_3 \\ \mathbf{G}_3 \end{matrix}(\boldsymbol{\xi}\otimes\boldsymbol{\xi}\otimes\boldsymbol{\xi}) + \cdots \tag{1.6}$$

Symbol $\otimes$ denotes the Kronecker vector product (tensor product) and the shorthand notation $\boldsymbol{\xi}^{\otimes k}$ will be used to denote the k-th power of vector $\boldsymbol{\xi}$ in the Kronecker product sense.

The procedure of producing the terms in the expansion above is carried out in an element-wise manner. Specifically, it is demonstrated for the case of matrix function **F** but can be straightforwardly generalized for $\boldsymbol{\psi}$ and **G**. First the *multi-index*, α, *notation* is introduced to simplify the expressions

$$\alpha \triangleq (\alpha_1, \alpha_2, \ldots, \alpha_N); \quad |\alpha| \triangleq \alpha_1 + \alpha_2 + \cdots + \alpha_N,$$
$$\alpha! \triangleq (\alpha_1!)(\alpha_2!)\cdots(\alpha_N!); \quad \alpha_1, \alpha_2, \ldots, \alpha_N \in \mathbb{N}.$$

In effect:

$$\begin{gathered} \mathbf{F}(\boldsymbol{\xi}) = \left[f_{ij}(\boldsymbol{\xi})\right], \quad 1 \le i, \quad j \le \mathbf{K}; \quad \boldsymbol{\xi} \triangleq \left[\begin{matrix} \xi_1 & \xi_2 & \ldots & \xi_N \end{matrix}\right]^T \\ \Downarrow \\ \mathbf{F}_\alpha\left(\boldsymbol{\xi}^{\otimes\alpha}\right) = \left[\mathbf{f}_{ij}^{(\alpha)} \cdot \boldsymbol{\xi}^{\otimes\alpha}\right], \quad 1 \le i, \quad j \le \mathbf{K}; \quad \boldsymbol{\xi}^{\otimes\alpha} \triangleq \left[\xi_1^{\alpha_1}\, \xi_2^{\alpha_2} \cdots \xi_N^{\alpha_N}\right]. \end{gathered} \tag{1.7}$$

In the above, the α-th derivative vector, $\mathbf{f}_{ij}^{(\alpha)}$ is defined by the following.

$$\mathbf{f}^{(\alpha)} \triangleq \left[\frac{1}{\alpha!}\partial^\alpha f\right]; \quad \partial^\alpha f \triangleq \frac{\partial^{|\alpha|} f}{\partial\xi_1^{\alpha_1}\, \partial\xi_2^{\alpha_2} \cdots \partial\xi_N^{\alpha_N}}. \tag{1.8}$$

One needs to match the expansion of the operator in case considered with respect to the Kronecker vector power $\mathbf{x}_1^{\otimes k}$; also, $\mathbf{K} = n_2$, and $\mathbf{N} = n_1$. Note that the partial derivatives, appearing in the operator expansion, are evaluated at $\mathbf{x}_1 = \mathbf{0}$ and some of the elements of $\mathbf{f}_{ij}^{(k)}$ may need to be set to zero to avoid repetitions, since an underlying assumption is Clairaut's Theorem [8] according to which the order of mixed derivatives can be interchanged. Special cases are $k = 0$ and $k = 1$:

$$\left.\begin{aligned} \mathbf{F}_0 = \mathbf{F}(\mathbf{0}) = \left[f_{ij}(\mathbf{x}_1 = \mathbf{0})\right], \quad \mathbf{x}_1^{\otimes 0} = \left[\begin{matrix} 1 & \ldots & 1 \end{matrix}\right]^T \\ \mathbf{F}_1(\mathbf{x}_1) = \left[\nabla f_{ij}(\mathbf{x}_1 = \mathbf{0}) \cdot \mathbf{x}_1\right], \quad \mathbf{x}_1^{\otimes 1} = \mathbf{x}_1 \end{aligned}\right\}; \quad 1 \le i, \quad j \le n_2. \tag{1.9}$$

Note that, as in the general case, the elements of $\mathbf{F}_1$ are scalar products of the Kronecker power vector for $k = 1$ times the element-wise partial derivative (grad) vector evaluated at $\mathbf{x}_1 = \mathbf{0}$.

By using the expansion for matrix functions $\mathbf{F}, \mathbf{G}$, the system equations are reformed as follows:

$$\begin{aligned} \dot{\mathbf{x}}_2 &= \mathbf{F}_0 \cdot \mathbf{x}_2 + \underbrace{\left[\begin{array}{cc} \mathbf{G}_0 & \mathbf{\Gamma}_0 \end{array}\right]}_{\mathbf{\Gamma}} \cdot \left[\begin{array}{c} \mathbf{u} \\ \mathbf{r} \end{array}\right] \\ \mathbf{y}_2 &= \mathbf{C}_2 \cdot \mathbf{x}_2 . \end{aligned} \tag{1.10}$$

In the above, auxiliary signal vector $\mathbf{r}$ is defined as follows:

$$\mathbf{r}(t) \triangleq \sum_{k=1}^{\infty} \left\{ \mathbf{F}_k \left(\mathbf{x}_1^{\otimes k} \right) \cdot \mathbf{x}_2 + \mathbf{G}_k \left(\mathbf{x}_1^{\otimes k} \right) \cdot \mathbf{u} \right\} . \tag{1.11}$$

Furthermore, it holds that:

$$\mathbf{\Gamma} \in \mathbb{R}^{n_2 \times (m_2 + n_2)}, \qquad \mathbf{\Gamma}_0 \in \mathbb{R}^{n_2 \times n_2}.$$

$\mathbf{\Gamma}_0$ is a square binary matrix (i.e., (0,1)-matrix) accounting for the fact that certain elements of signal vector $\mathbf{r}$ may be identically zero.

In the sequel, with respect to the BP segment of our original equations, systems of form (1.10) will be considered with the additional assumption that the following transfer function matrix is BP around a carrier frequency ω_c:

$$\mathbf{H}_2(s) = \mathbf{C}_2 \cdot (s\mathbf{I} - \mathbf{F}_0)^{-1} \cdot \mathbf{\Gamma} . \tag{1.12}$$

The coupling of such BP systems to LP systems will be investigated with the additional assumption that the following transfer function matrix is LP:

$$\mathbf{H}_1(s) = (s\mathbf{I} - \mathbf{A}_{LP})^{-1} . \tag{1.13}$$

The analysis is presented employing the mathematical apparatus to establish the equivalent low-pass model. This, however, requires the Hilbert Transform which is briefly presented next.

1.3 THE HILBERT TRANSFORM AND THE COMPLEX ENVELOPE OF A BP SIGNAL

The Hilbert transform allows analyzing phase selective and spectrally decoupled systems which interact through amplitude or phase modulation. The Hilbert transform, in contrast to the Laplace and Fourier transforms, does not establish a new domain but both its input and output are in the time domain. In literature, the Hilbert transform is often seen as a phase shift applied to a scalar or vector signal. As seen later on, superimposing the Hilbert transform of a BP signal with the signal itself results in the complex pre-envelope; the complex pre-envelope in turn can then be shifted down to baseband to yield the LP complex envelope which is essentially the information carried by the signal.

A more formal definition as an integral transform is given below for a signal vector:

$$\hat{\mathbf{x}}(t) \triangleq \frac{1}{\pi} \int_{-\infty}^{+\infty} \frac{1}{t - t_1} \mathbf{x}(t_1)\, dt_1 = \frac{1}{\pi t} * \mathbf{x}(t) \tag{1.14}$$
$$\Updownarrow$$
$$\hat{\mathbf{x}}(f) = -j \operatorname{sgn}(f)\, \mathbf{x}(f).$$

In the equation above, the star (*) stands for the convolution operator. Specifically, the definition of the convolution operator, for two real or complex functions of one, real variable, is:

$$\begin{aligned} \phi(\chi) &= \phi_1(\chi) * \phi_2(\chi) \\ &\triangleq \int_{-\infty}^{+\infty} \phi_1(\xi)\phi_2(\chi - \xi)d\xi \\ &= \int_{-\infty}^{+\infty} \phi_2(\xi)\phi_1(\chi - \xi)d\xi = \phi_2(\chi) * \phi_1(\chi). \end{aligned} \tag{1.15}$$

The sign function is defined below:

$$\operatorname{sgn}(\xi) = \begin{cases} +1, & \xi > 0 \\ 0, & \xi = 0 \\ -1, & \xi < 0. \end{cases} \tag{1.16}$$

The Fourier transform to be used in this text is the bilateral Fourier integral given below for a signal vector:

$$\mathbf{x}(f) \triangleq \int_{-\infty}^{+\infty} \mathbf{x}(t) \exp(-j2\pi f t)\, dt \tag{1.17}$$
$$\Updownarrow$$
$$\mathbf{x}(t) = \int_{-\infty}^{+\infty} \mathbf{x}(f) \exp(j2\pi f t)\, df.$$

The Hilbert transform is linear and has an inverse transform, defined as follows:

$$\mathbf{x}(t) \triangleq -\frac{1}{\pi} \int_{-\infty}^{+\infty} \frac{1}{t - t_1} \hat{\mathbf{x}}(t_1) dt_1. \tag{1.18}$$

Two important properties make the Hilbert transform extremely useful in the analysis of BP signals and systems especially when amplitude modulation is involved. The first one concerns the product of a BP and a LP signal; the scalar case is given below:

$$\left.\begin{array}{l}\chi(t)=\chi_1(t)\,\chi_2(t)\\ \chi_1(t)\in\mathbb{R}\ \text{ is }\ LP\\ \chi_2(t)\in\mathbb{R}\ \text{ is }\ BP\end{array}\right\}\Rightarrow\hat{\chi}(t)=\chi_1(t)\,\hat{\chi}_2(t)\,. \tag{1.19}$$

The sinusoidal signals are linked as phase shifts by the Hilbert transform:

$$\begin{aligned}\chi(t)&=\cos(2\pi f_0 t)\Rightarrow\hat{\chi}(t)=\sin(2\pi f_0 t)\\ \chi(t)&=\sin(2\pi f_0 t)\Rightarrow\hat{\chi}(t)=-\cos(2\pi f_0 t)\,.\end{aligned} \tag{1.20}$$

Using the Hilbert transform, the pre-envelope, $\mathbf{x}_+(t)$, of a signal vector, $\mathbf{x}(t)$, is defined as a signal vector with elements in $\mathbb{C}$ as follows:

$$\begin{gathered}\mathbf{x}_+(t)\triangleq\mathbf{x}(t)+j\hat{\mathbf{x}}(t)\\ \Updownarrow\\ \mathbf{x}_+(f)=\mathbf{x}(f)+j\left(-j\,\mathrm{sgn}(f)\mathbf{x}(f)\right)=\begin{cases}2\mathbf{x}(f), & f>0\\ \mathbf{x}(f), & f=0\\ 0, & f<0.\end{cases}\end{gathered} \tag{1.21}$$

As can be seen the pre-envelope is a complex signal vector allowed to have nonzero spectrum only in non-negative frequencies. Using the pre-envelope, $\mathbf{x}_+(t)$, the complex envelope, $\tilde{\mathbf{x}}(t)$, is defined for a BP signal vector, $\mathbf{x}(t)$, with carrier frequency $\omega_c=2\pi f_c$:

$$\begin{gathered}\mathbf{x}_+(t)=\tilde{\mathbf{x}}(t)\exp(j2\pi f_c t)\\ \Updownarrow\\ \tilde{\mathbf{x}}(t)\triangleq\mathbf{x}_+(t)\exp(-j2\pi f_c t)\\ \Updownarrow\\ \tilde{\mathbf{x}}(f)=\mathbf{x}_+(f-f_c)=2\mathbf{x}(f-f_c)\,.\end{gathered} \tag{1.22}$$

The last one of the above relationships in combination with the fact that signal $\mathbf{x}(t)$ is BP with carrier frequency $\omega_c=2\pi f_c$ leads to the conclusion that the complex envelope of a BP signal is an LP signal.

The complex envelope may be decomposed to a real and an imaginary component:

$$\tilde{\mathbf{x}}(t)=\mathbf{x}_C(t)+j\mathbf{x}_S(t). \tag{1.23}$$

In telecommunications literature the real component, $\mathbf{x}_C(t)$, is referred to as the in-phase (or I for short) component and the imaginary component, $\mathbf{x}_S(t)$, is referred to as the quadrature (or Q for short) component. Clearly, the I and Q components of the complex envelope are mutually orthogonal and preserve the complete information content of the BP signal from which they are generated. Furthermore, as can be seen from the first one of Equations (1.22), the complex envelope is a generalization of the concept of amplitude modulation applied to the generalized imaginary exponential carrier signal $\exp(j\omega_c t)$. The following algebraic manipulation supports this claim:

$$\begin{gathered}\mathbf{x}(t) = \operatorname{Re}[\mathbf{x}_+(t)] = \operatorname{Re}[\tilde{\mathbf{x}}(t)\exp(j2\pi f_c t)] \\ \Downarrow \\ \mathbf{x}(t) = \mathbf{x}_C(t)\cos(\omega_c t) - \mathbf{x}_S(t)\sin(\omega_c t).\end{gathered} \tag{1.24}$$

The above clearly demonstrates that the original (real) signal vector is generated by amplitude modulation of the carrier signal $\cos(\omega_c t)$ by the I component and amplitude modulation of the carrier signal $\sin(\omega_c t)$ by the Q component. Note that the two carriers, although demonstrating the same frequency, are mutually orthogonal.

1.4 EQUIVALENT LOW-PASS MODEL OF A SYSTEM WITH STATE-SPACE MODULATION

We now return to the coupled oscillators, Equations (1.1) and (1.2), by taking into account that the transfer function matrix is BP for one segment of the system while the other one LP. Then, *an equivalent, exclusively LP system model may be obtained, by using the complex envelope of the BP signal vector* $\mathbf{x}_2(t)$.

At first, due to the assumption state vector $\mathbf{x}_1(t)$ is LP, as it is generated by an LP system. Indeed, assuming zero initial conditions, the state vector is given in the frequency domain by the following relation:

$$\mathbf{x}_1(\omega) = \mathbf{H}_1(\omega) \cdot \mathbf{d}_1(\omega). \tag{1.25}$$

In the above, the auxiliary signal $\mathbf{d}_1(t)$ is defined as follows:

$$\mathbf{d}_1(t) = \boldsymbol{\psi}(\mathbf{y}_2(t)) + \mathbf{d}(t). \tag{1.26}$$

Therefore, if $\mathbf{d}_1(t)$ is assumed to be a random signal of white Gaussian noise content, i.e., possessing power spectral density constant over all frequencies, $\mathbf{x}_1(t)$ will comply to the low-pass requirement.

The next step is to observe that an arbitrary Kronecker power expansion of an LP signal vector is also LP with possibly larger BW. Therefore, the elements of matrices $\mathbf{F}$, $\mathbf{G}$ if viewed as scalar signals are LP, at least in the case that the expansions in Equation (1.6) obtain a finite number of terms. Property (1.19) guarantees, then, that the following holds:

$$\begin{gathered}\dot{\mathbf{x}}_2 = \mathbf{F}(\mathbf{x}_1) \cdot \mathbf{x}_2 + \mathbf{G}(\mathbf{x}_1) \cdot \mathbf{u} \\ \Downarrow \\ \dot{\hat{\mathbf{x}}}_2 = \mathbf{F}(\mathbf{x}_1) \cdot \hat{\mathbf{x}}_2 + \mathbf{G}(\mathbf{x}_1) \cdot \hat{\mathbf{u}}.\end{gathered} \tag{1.27}$$

In the above, the following fact for the time derivative (denoted by a dot placed above the signal) of a signal vector has been used:

$$\hat{\dot{\mathbf{x}}}(t) = \dot{\hat{\mathbf{x}}}(t). \tag{1.28}$$

This is a direct consequence of the linearity property of the Hilbert transform. If the second equation in (1.27) is multiplied by j and then added to the first, the following dynamic equation for the pre-envelope, $\mathbf{x}_{2+}$, of the state vector signal $\mathbf{x}_2$ is obtained:

$$\dot{\mathbf{x}}_{2+} = \mathbf{F}(\mathbf{x}_1) \cdot \mathbf{x}_{2+} + \mathbf{G}(\mathbf{x}_1) \cdot \mathbf{u}_{+}. \tag{1.29}$$

Because of the assumption that the transfer function in Equation (1.12) is BP with carrier frequency $\omega_c = 2\pi f_c$ the pre-envelope $\mathbf{x}_{2+}$, of the state vector signal $\mathbf{x}_2$ may be expressed as a product between a modulating complex envelope factor, $\tilde{\mathbf{x}}_2$, and an imaginary scalar exponential signal acting as a generalized sinusoidal carrier signal:

$$\begin{gathered} \mathbf{x}_{2+}(t) = \tilde{\mathbf{x}}_2(t)\exp(j\omega_c t) \\ \Downarrow \\ \dot{\mathbf{x}}_{2+}(t) = \dot{\tilde{\mathbf{x}}}_2(t)\exp(j\omega_c t) + j\omega_c\tilde{\mathbf{x}}_2(t)\exp(j\omega_c t). \end{gathered} \tag{1.30}$$

If the input signal vector $\mathbf{u}$ is assumed tuned to the BP system, then it may be expressed similarly as follows:

$$\mathbf{u}_{+}(t) = \tilde{\mathbf{u}}(t)\exp(j\omega_c t). \tag{1.31}$$

By substituting the equations above in Equation (1.29) the following relationship is finally obtained for the complex envelope of the BP system's state vector:

$$\dot{\tilde{\mathbf{x}}}_2 = [\mathbf{F}(\mathbf{x}_1) - j\omega_c\mathbf{I}] \cdot \tilde{\mathbf{x}}_2 + \mathbf{G}(\mathbf{x}_1) \cdot \tilde{\mathbf{u}}. \tag{1.32}$$

If a similar expression for the output vector, $\mathbf{y}_2$ is adopted then one obtains the following expression for its complex envelope $\tilde{\mathbf{y}}_2$; indeed, if the first Equation (1.30) is substituted in the output equation of the BP subsystem:

$$\left.\begin{aligned} \mathbf{y}_{2+}(t) &= \tilde{\mathbf{y}}_2(t)\exp(j\omega_c t) \\ \mathbf{y}_2(t) &= \mathrm{Re}\{\mathbf{y}_{2+}(t)\} \end{aligned}\right\} \Rightarrow \tilde{\mathbf{y}}_2 = \mathbf{C}_2 \cdot \tilde{\mathbf{x}}_2. \tag{1.33}$$

By using the above, one can rewrite output equation of the BP subsystem so that it contains the I and Q components of $\tilde{\mathbf{y}}_2$, $\mathbf{y}_{2C}$, and $\mathbf{y}_{2S}$, respectively. This is done by employing directly the expression for $\boldsymbol{\psi}$ of Equation (1.6) in Equation (1.24). However, further simplification is possible by exploiting the assumption that transfer function (1.13) is LP in order to eliminate high-frequency terms, i.e., terms including a "carrier" factor of the form $\exp(\pm j\omega_c t)$, $k = 1, 2, 3, \ldots$. Such factors appear as a direct consequence of the fact that the coupling term in the LP subsystem's dynamic equation is the nonlinear function $\boldsymbol{\psi}$. Then, terms with factors of the form $\exp(\pm j\omega_c t)$, $k = 1, 2, 3, \ldots$ may be neglected on the basis of the spectral decoupling between the LP and the BP system. This is translated to the requirement that the LP system's BW is sufficiently smaller than the BP system's carrier frequency $\omega_c = 2\pi f_c$.

By rewriting Equations (1.6) and (1.7) in the case of multivariable vector function $\boldsymbol{\psi}(\mathbf{y}_2)$ one can obtain that:

$$\boldsymbol{\psi}(\mathbf{y}_2) = \boldsymbol{\psi}(\mathbf{y}_2 = \mathbf{0}) + \left[\sum_{k=1}^{\infty} \boldsymbol{\psi}_i^{(k)} \cdot \mathbf{y}_2^{\otimes k}\right], \quad 1 \le i \le n_1. \tag{1.34}$$

An LP contribution from the above is possible only if k is even. Indeed, an LP contribution is this part of each term in the above expansion which is not multiplied by a carrier factor $\exp(\pm j\omega_c t)$, $k = 1, 2, 3, \ldots$. Therefore, despite the 0-th term, which obviously participates in the LP part of the signal vector in the equation above for $\boldsymbol{\psi}$, all terms for strictly positive k contain a component multiplied by $\exp(\pm j\omega_c t)$, $k = 1, 2, 3, \ldots$. However, when k is odd only this component is present, e.g., for $k = 1$. On the other hand, when k is even, except the carrier-multiplied component, there exists a carrier-free one, too. This is made clearer with an example, e.g., for $k = 2$.

In this case:

$$\begin{aligned}
\mathbf{y}_2^{\otimes 2} &= \left(\mathrm{Re}\left[\tilde{\mathbf{y}}_2 e^{j\omega_c t}\right]\right)^{\otimes 2} = \left(\frac{\tilde{\mathbf{y}}_2 e^{j\omega_c t} + \tilde{\mathbf{y}}_2^* e^{-j\omega_c t}}{2}\right)^{\otimes 2} \\
&= \frac{\tilde{\mathbf{y}}_2 \otimes \tilde{\mathbf{y}}_2^*}{2} + \left(\frac{\tilde{\mathbf{y}}_2}{2}\right)^{\otimes 2} e^{j2\omega_c t} + \left(\frac{\tilde{\mathbf{y}}_2^*}{2}\right)^{\otimes 2} e^{-j2\omega_c t} \\
&= \frac{\mathbf{y}_{2C}^{\otimes 2} + \mathbf{y}_{2S}^{\otimes 2}}{2} + \left(\frac{\tilde{\mathbf{y}}_2}{2}\right)^{\otimes 2} e^{j2\omega_c t} + \left(\frac{\tilde{\mathbf{y}}_2^*}{2}\right)^{\otimes 2} e^{-j2\omega_c t}.
\end{aligned} \tag{1.35}$$

In the above, only the first term is an LP one as the other two contain a carrier factor.

In conclusion, Equations (1.19) and (1.20) of coupled oscillators, with LP and BP transfer function matrices as in Equations (1.31) and (1.30), respectively, *can be reduced to the following LP equivalent system of equations:*

$$\dot{\mathbf{x}}_1 = \mathbf{A}_{LP} \cdot \mathbf{x}_1 + \boldsymbol{\psi}_{LP}(\tilde{\mathbf{y}}_2) + \mathbf{d} \tag{1.36}$$

$$\begin{aligned}
\dot{\tilde{\mathbf{x}}}_2 &= \left[\mathbf{F}(\mathbf{x}_1) - j\omega_c \mathbf{I}\right] \cdot \tilde{\mathbf{x}}_2 + \mathbf{G}(\mathbf{x}_1) \cdot \tilde{\mathbf{u}} \\
\tilde{\mathbf{y}}_2 &= \mathbf{C}_2 \cdot \tilde{\mathbf{x}}_2.
\end{aligned} \tag{1.37}$$

In the above, all signals in the BP subsystem have been substituted by their LP complex envelopes, e.g., $\mathbf{x}_2(t)$ by $\tilde{\mathbf{x}}_2(t) = \mathbf{x}_{2C}(t) + j\mathbf{x}_{2S}(t)$. Furthermore, in the LP subsystem equation, multivariable vector function $\boldsymbol{\psi}(\mathbf{y}_2)$ has been substituted by the LP one $\boldsymbol{\psi}_{LP}(\tilde{\mathbf{y}}_2) = \boldsymbol{\psi}_{LP}(\mathbf{y}_{2C}, \mathbf{y}_{2S})$, which is produced by omitting from the original all odd terms and the modulated carrier parts of the even terms. The main benefit in using the description of Equations (1.36) and (1.37) instead of the original ones in Equations (1.1) and (1.2) is that, because it is LP but otherwise grasps all the essential dynamics of the dynamical system at hand, the time step for the integration of the dynamic equations may be set to a substantially smaller value than in the original one. For example, in typical mechatronic applications, as the one presented as example later in this text, the BW of both the LP and the BP system is commonly in the order of magnitude of 10 Hz. However, the carrier frequency may be 1 kHz or even higher. Therefore, the integration step may be increased at least two orders of magnitude; such a possibility makes investigations using numerical simulation much easier. Another benefit is that by using the equivalent LP system the carrier frequency, which does not play such a crucial role in the understanding of the dynamics, comes into the analysis simply as a selectable parameter. In effect, as far as the main assumptions are satisfied the selection of the carrier frequency does not affect any significant conclusions for the behavior of the system at hand.

CHAPTER 2

Exact Feedback Linearization

2.1 CONCEPT AND MOTIVATION

Feedback linearization is a common approach used in controlling nonlinear systems. The approach involves coming up with a transformation of the nonlinear system into an equivalent linear system through a change of variables and a suitable control input.

Feedback linearization may be applied to nonlinear systems of the following affine form [9]:

$$\begin{aligned} \dot{\mathbf{x}} &= \mathbf{f}(\mathbf{x}) + \mathbf{g}(\mathbf{x}) \cdot \mathbf{u} \\ \mathbf{y} &= \mathbf{h}(\mathbf{x}). \end{aligned} \tag{2.1}$$

The various terms entering the above equations are defined as follows:

$$\begin{aligned} &\mathbf{x} \in \boldsymbol{D},\ \mathbf{u} \in \mathbb{R}^m,\ \mathbf{y} \in \mathbb{R}^p,\ \boldsymbol{D} \subset \mathbb{R}^n \\ &\mathbf{f}: \mathbf{x} \in \boldsymbol{D} \rightarrow \mathbf{f}(\mathbf{x}) \in \mathbb{R}^n \\ &\mathbf{g}: \mathbf{x} \in \boldsymbol{D} \rightarrow \mathbf{g}(\mathbf{x}) \in \mathbb{R}^{n \times m} \\ &\mathbf{h}: \mathbf{x} \in \boldsymbol{D} \rightarrow \mathbf{h}(\mathbf{x}) \in \mathbb{R}^p. \end{aligned}$$

Domain $\boldsymbol{D}$ is formally required to contain the origin. Then, a control input $\mathbf{u}$ and a change of variables $\boldsymbol{T}$ of the following form are sought after:

$$\mathbf{u} = \boldsymbol{\alpha}(\mathbf{x}) + \boldsymbol{\beta}(\mathbf{x}) \cdot \mathbf{v} \tag{2.2}$$

$$\mathbf{z} = \boldsymbol{T}(\mathbf{x}). \tag{2.3}$$

We typically assume that the dimension of vector $\mathbf{v}$ is the same with that of $\mathbf{u}$ for simplicity. The objective is to render a linear input-state or input-output map between the new input $\mathbf{v}$ and the state vector or the output. An outer-loop control strategy for the resulting linear control system can then be applied [9–11]. If the answer to this question is positive, we can induce linear behavior in nonlinear systems and apply the large number of tools and the well-established theory of linear control to develop stabilizing controllers.

Note that the original system introduced in the previous chapter is a special case of the one in Equation (2.1) especially when undisturbed ($\mathbf{d} = \mathbf{0}$). Indeed, the dynamic equation can be rewritten as follows:

$$\dot{\mathbf{x}} = \underbrace{\begin{bmatrix} \mathbf{A}_{LP} & \mathbf{0} \\ \mathbf{0} & \mathbf{F}(\mathbf{x}_1) \end{bmatrix} \mathbf{x} + \begin{bmatrix} \boldsymbol{\psi}(\mathbf{C}_2\mathbf{x}_2) \\ \mathbf{0} \end{bmatrix}}_{\mathrm{f(x)}} + \underbrace{\begin{bmatrix} \mathbf{0} \\ \mathbf{G}(\mathbf{x}_1) \end{bmatrix}}_{\mathrm{g(x)}} \mathbf{u}; \quad \mathbf{x} = \begin{bmatrix} \mathbf{x}_1 \\ \mathbf{x}_2 \end{bmatrix}. \tag{2.4}$$

Also, note that the equivalent LP system introduced in the previous chapter is a special case of the one in Equation (2.1) especially when undisturbed ($\mathbf{d} = \mathbf{0}$). Indeed the dynamic equation can be rewritten as follows:

$$\dot{\mathbf{x}} = \underbrace{\begin{bmatrix} \mathbf{A}_{LP} & \mathbf{0} \\ \mathbf{0} & \mathbf{F}(\mathbf{x}_1) - j\omega_c \mathbf{I} \end{bmatrix} \mathbf{x} + \begin{bmatrix} \boldsymbol{\psi}_{LP}(\mathbf{C}_2 \tilde{\mathbf{x}}_2) \\ \mathbf{0} \end{bmatrix}}_{\mathbf{f}(\mathbf{x})} + \underbrace{\begin{bmatrix} \mathbf{0} \\ \mathbf{G}(\mathbf{x}_1) \end{bmatrix}}_{\mathbf{g}(\mathbf{x})} \tilde{\mathbf{u}}; \quad \mathbf{x} = \begin{bmatrix} \mathbf{x}_1 \\ \tilde{\mathbf{x}}_2 \end{bmatrix}. \tag{2.5}$$

2.2 INPUT-STATE LINEARIZATION

Clearly, we should not expect to be able to cancel nonlinearities in every nonlinear system. There must be a certain structural property of the system that allows us to perform such cancellation. Therefore, the ability to use feedback to convert a nonlinear state equation into a controllable linear state equation by canceling nonlinearities requires the nonlinear space equation to have the following structure:

$$\dot{\mathbf{x}} = \mathbf{A}\mathbf{x} + \mathbf{B}\boldsymbol{\beta}^{-1}(\mathbf{x})\left[\mathbf{u} - \boldsymbol{\alpha}(\mathbf{x})\right]. \tag{2.6}$$

In the previous equation matrix $\mathbf{A}$ is $n \times n$, matrix $\mathbf{B}$ is $n \times m$, and the pair $(\mathbf{A}, \mathbf{B})$ is *controllable*. Vector function $\boldsymbol{\alpha}$ and matrix function $\boldsymbol{\beta}^{-1}$ need to be defined on domain $\boldsymbol{D}$, which as before is formally required to contain the origin.

$$\begin{aligned} &\boldsymbol{\alpha} : \mathbf{x} \in \boldsymbol{D} \subset \mathbb{R}^n \to \boldsymbol{\alpha}(\mathbf{x}) \in \mathbb{R}^m \\ &\boldsymbol{\beta}^{-1} : \mathbf{x} \in \boldsymbol{D} \subset \mathbb{R}^n \to \boldsymbol{\beta}^{-1}(\mathbf{x}) \in \mathbb{R}^{m \times m}. \end{aligned}$$

Furthermore, matrix $\boldsymbol{\beta}^{-1}$, which as the notation implies is the inverse of $\boldsymbol{\beta}(\mathbf{x})$ appearing in Equation (2.2), must be nonsingular over the entirety of domain $\boldsymbol{D}$. Then, if state feedback as in Equation (2.2) is applied to the system, the following linear dynamic equation is obtained:

$$\dot{\mathbf{x}} = \mathbf{A}\mathbf{x} + \mathbf{B}\mathbf{v}. \tag{2.7}$$

The system can then be treated as linear. For example, a *stabilizing controller* of the form $\mathbf{v} = -\mathbf{K}\mathbf{x}$ can be introduced by choosing state feedback matrix gain so that matrix $(\mathbf{A} - \mathbf{B}\mathbf{K})$ is *Hurwitz stable*. In effect, the overall nonlinear stabilizing state feedback control law is given by the following:

$$\mathbf{u} = \boldsymbol{\alpha}(\mathbf{x}) - \boldsymbol{\beta}(\mathbf{x})\mathbf{K}\mathbf{v}. \tag{2.8}$$

In the case, however, that the nonlinear state equation does not have the required structure as in Equation (2.6) the system may still be linearizable. This is because even if the state equation does not have the required structure for one choice of variables, it might do so for another choice. We proceed now to find a transformation of the state variables so that the transformed state equation has the required structure.

2.3 THE TRANSFORMATION MATRIX

Consider a nonlinear system which is affine per the control input, i.e., of the following form:

$$\dot{\mathbf{x}} = \mathbf{f}(\mathbf{x}) + \mathbf{g}(\mathbf{x}) \cdot \mathbf{u}. \tag{2.9}$$

Assume that $\mathbf{f} : \mathbf{x} \in \boldsymbol{D} \to \mathbf{f}(\mathbf{x}) \in \mathbb{R}^n$ and $\mathbf{g} : \mathbf{x} \in \boldsymbol{D} \to \mathbf{g}(\mathbf{x}) \in \mathbb{R}^{n \times m}$ are sufficiently smooth on a domain $\boldsymbol{D}$. Then, the system is *input-state linearizable* (or feedback linearizable) if there exists a *diffeomorphism* $\boldsymbol{T} : \boldsymbol{D} \to \mathbb{R}^n$ such that domain $\boldsymbol{D}_z = \boldsymbol{T}(\boldsymbol{D})$ contains the origin and the change of variables $\mathbf{z} = \boldsymbol{T}(\mathbf{x})$ transforms the system state equation to the form:

$$\dot{\mathbf{z}} = \mathbf{A}\mathbf{z} + \mathbf{B}\boldsymbol{\beta}^{-1}(\mathbf{z})\left[\mathbf{u} - \boldsymbol{\alpha}(\mathbf{z})\right]. \tag{2.10}$$

In the above, $(\mathbf{A}, \mathbf{B})$ is *controllable* and $\boldsymbol{\beta}^{-1}(\mathbf{z})$ is nonsingular at least in domain $\boldsymbol{D}$. It is noted here that a function is called smooth if it is continuous and if all of its derivatives of any order are also continuous. Also, a transformation like $\boldsymbol{T}$ is called a diffeomorphism if it is smooth and its inverse exists and is smooth as well.

It is noted here that if certain output variables are of interest, e.g., in tracking control, linearizing the state does not necessarily imply that the output is linearized as well. We will revisit this topic soon since solving such problems may still be complicated even after input-state linearization has been successfully applied.

Returning to the input-state linearization process for now, a straightforward process is derived to obtain equations to determine $\boldsymbol{T}$. We have that:

$$\left.\begin{array}{l} \dot{\mathbf{x}} = \mathbf{f}(\mathbf{x}) + \mathbf{g}(\mathbf{x}) \cdot \mathbf{u} \\ \mathbf{z} = \boldsymbol{T}(\mathbf{x}) \end{array}\right\} \Rightarrow \dot{\mathbf{z}} = \frac{\partial \boldsymbol{T}}{\partial \mathbf{x}} \cdot \dot{\mathbf{x}} = \frac{\partial \boldsymbol{T}}{\partial \mathbf{x}} \mathbf{f}(\mathbf{x}) + \frac{\partial \boldsymbol{T}}{\partial \mathbf{x}} \mathbf{g}(\mathbf{x}) \mathbf{u}$$
$$\dot{\mathbf{z}} = \mathbf{A}\mathbf{z} + \mathbf{B}\boldsymbol{\beta}^{-1}(\mathbf{z})\left[\mathbf{u} - \boldsymbol{\alpha}(\mathbf{z})\right].$$

Then, one derives that the following partial differential equations are required to hold for $\boldsymbol{T}$:

$$\begin{aligned} \frac{\partial \boldsymbol{T}}{\partial \mathbf{x}} \cdot \mathbf{f}(\mathbf{x}) &= \mathbf{A} \cdot \boldsymbol{T}(\mathbf{x}) - \mathbf{B} \cdot \boldsymbol{\beta}^{-1}(\boldsymbol{T}(\mathbf{x})) \cdot \boldsymbol{\alpha}(\boldsymbol{T}(\mathbf{x})) \\ \frac{\partial \boldsymbol{T}}{\partial \mathbf{x}} \cdot \mathbf{g}(\mathbf{x}) &= \mathbf{B} \cdot \boldsymbol{\beta}^{-1}(\boldsymbol{T}(\mathbf{x})). \end{aligned} \tag{2.11}$$

Evidently, the expressions above are quite complicated and hard to process further. However, a further result can be achieved, if constant matrices $\mathbf{A}$ and $\boldsymbol{B}$ are assumed to be in the *controllable canonical form*. Indeed, transformation $\boldsymbol{T}$ is not unique. One can consider a further transformation of $\mathbf{z}$ to a new state vector $\boldsymbol{\xi}$ of the same dimension through a transformation assumed to be linear; in effect, a constant transformation matrix $\boldsymbol{\Xi}$ is employed. Evidently, $\boldsymbol{\Xi}$ needs to be nonsingular, i.e., Ξ^{-1} to exist.

$$\begin{gathered}\boldsymbol{\xi} = \Xi \cdot \mathbf{z}, \quad \frac{d}{dt}\boldsymbol{\xi} = \Xi \cdot \dot{\mathbf{z}} \\ \Updownarrow \\ \mathbf{z} = \Xi^{-1} \cdot \boldsymbol{\xi}, \quad \dot{\mathbf{z}} = \Xi^{-1} \cdot \frac{d}{dt}\boldsymbol{\xi}.\end{gathered} \tag{2.12}$$

Then, the state equation in terms of $\mathbf{z}$ obtains the following form in terms of the new state vector $\boldsymbol{\xi}$. The two state vectors are evidently of the same dimension n:

$$\dot{\mathbf{z}} = \mathbf{A}\mathbf{z} + \mathbf{B}\boldsymbol{\beta}^{-1}(\mathbf{z})\left[\mathbf{u} - \boldsymbol{\alpha}(\mathbf{z})\right]$$

$$\begin{gathered}\Updownarrow \\ \Xi^{-1} \cdot \frac{d}{dt}\boldsymbol{\xi} = \mathbf{A}\Xi^{-1} \cdot \boldsymbol{\xi} + \mathbf{B} \cdot \boldsymbol{\beta}^{-1}\left(\Xi^{-1}\boldsymbol{\xi}\right) \cdot \left[\mathbf{u} - \boldsymbol{\alpha}\left(\Xi^{-1}\boldsymbol{\xi}\right)\right] \\ \Updownarrow \\ \frac{d}{dt}\boldsymbol{\xi} = \underbrace{\Xi\mathbf{A}\Xi^{-1}}_{\mathbf{A}_\Xi} \cdot \boldsymbol{\xi} + \underbrace{\Xi\mathbf{B}}_{\mathbf{B}_\Xi} \cdot \boldsymbol{\beta}^{-1}\left(\Xi^{-1}\boldsymbol{\xi}\right) \cdot \left[\mathbf{u} - \boldsymbol{\alpha}\left(\Xi^{-1}\boldsymbol{\xi}\right)\right] \\ \Updownarrow \\ \frac{d}{dt}\boldsymbol{\xi} = \mathbf{A}_\Xi \boldsymbol{\xi} + \mathbf{B}_\Xi \boldsymbol{\beta}^{-1}\left(\Xi^{-1}\boldsymbol{\xi}\right)\left[\mathbf{u} - \boldsymbol{\alpha}\left(\Xi^{-1}\boldsymbol{\xi}\right)\right].\end{gathered} \tag{2.13}$$

Matrices $\mathbf{A}_\Xi$ and $\boldsymbol{B}_\Xi$ in the above are given by the following:

$$\begin{aligned}\mathbf{A}_\Xi &= \Xi \cdot \mathbf{A} \cdot \Xi^{-1} \\ \mathbf{B}_\Xi &= \Xi \cdot \mathbf{B}.\end{aligned} \tag{2.14}$$

The linear transformation Ξ is employed so matrices $\mathbf{A}_\Xi$ and $\mathbf{B}_\Xi$ can be put in the *controllable companion canonical form*. Without major loss of generality, the *single-input* $(m = 1)$ case will be considered from now on. Then, the following forms for matrices $\mathbf{A}_\Xi$ and $\mathbf{B}_\Xi$ are assumed:

$$\mathbf{B}_\Xi \equiv \mathrm{b_I} \triangleq \begin{bmatrix} 0 \\ 0 \\ \vdots \\ 1 \end{bmatrix}, \quad \mathbf{A}_\Xi \equiv \begin{bmatrix} 0 & 1 & \dots & 0 \\ 0 & \vdots & \ddots & \vdots \\ \vdots & 0 & \dots & 1 \\ -a_0 & -a_1 & \dots & -a_{n-1} \end{bmatrix} = \mathrm{A_{OI}} + \mathrm{b_I} \cdot \underline{\mathrm{a}}. \tag{2.15}$$

In the above:

$$\mathrm{A_{OI}} \triangleq \begin{bmatrix} 0 & 1 & \dots & 0 \\ 0 & \vdots & \ddots & \vdots \\ \vdots & 0 & \dots & 1 \\ 0 & 0 & \dots & 0 \end{bmatrix} = \begin{bmatrix} 0 & & & \\ 0 & & \mathbf{I}_{n-1} & \\ \vdots & & & \\ 0 & 0 & \dots & 0 \end{bmatrix}, \quad \underline{\mathrm{a}} \triangleq \begin{bmatrix} -a_0 & -a_1 & \dots & -a_{n-1} \end{bmatrix}.$$

Notice that $n \times n$ matrix $\mathrm{A_{OI}}$ and $n \times 1$ (column) vector $\mathrm{b_I}$ are standard in the sense that they are not only constant but also independent of the system considered. In contrast, $1 \times n$ row vector $\underline{a}$ reflects specifications for the linearized system. The freedom of selection of $\underline{a}$ is very useful when designing a controller for the linearized system to shape an outer loop that meets further specifications. For now it will be considered as a known vector, though.

We now return to Equations (2.11), the first of which can be rewritten as follows given that matrices $\mathbf{A}$ and $\mathbf{B}$ are assumed to be such so that the resulting system in state vector $\boldsymbol{\xi}$ is in the controllable companion canonical form. To see how this can be derived we rewrite Equation (2.10) as follows:

$$\begin{gathered}\dot{\mathbf{z}} = \left(\mathrm{A_{OI}} + \mathrm{b_I} \cdot \underline{\mathrm{a}}\right) \mathbf{z} + \mathrm{b_I} \cdot \frac{u - \alpha\left(\mathbf{z}\right)}{\beta\left(\mathbf{z}\right)} = \mathrm{A_{OI}}\mathbf{z} + \mathrm{b_I}\left[\underline{\mathrm{a}} \cdot \mathbf{z}\right] + \mathrm{b_I} \cdot \frac{u - \alpha\left(\mathbf{z}\right)}{\beta\left(\mathbf{z}\right)} \\ \Downarrow \\ \dot{\mathbf{z}} = \mathrm{A_{OI}}\mathbf{z} + b_{\mathrm{I}} \cdot \beta^{-1}\left(\mathbf{z}\right) \cdot \left[u - \left(\alpha\left(\mathbf{z}\right) - \beta\left(\mathbf{z}\right) \cdot \left[\underline{\mathrm{a}} \cdot \mathbf{z}\right]\right)\right].\end{gathered} \tag{2.16}$$

Also, notice that without major loss of generality and in the interest of brevity, the system has been assumed to be single-input, so functions $\boldsymbol{\alpha}(\mathbf{z})$, $\boldsymbol{\beta}(\mathbf{z})$, and $\boldsymbol{\beta}^{-1}(\mathbf{z})$, as well as $\mathbf{g}(\mathbf{x})$ in the original system, become scalar, i.e., $\alpha(\mathbf{z})$, $\beta(\mathbf{z})$, and $\beta^{-1}(\mathbf{z})$, as well as $\mathrm{g}(\mathbf{x})$, respectively. The choice $\mathbf{A} = \mathrm{A_{OI}} + \mathrm{b_I}\,\underline{\mathrm{a}}$ and $\mathbf{B} = \mathrm{b_I}$ not only simplifies the derivations but also leads to a linear system in controllable companion canonical form; the latter is pretty handy as shown later for the design of outer-loop controllers to improve performance of the linearized system. Indeed, if in the latter equation appearing in (2.16) we set the control $u = \beta(\mathbf{z})v + \beta(\mathbf{z})$, we obtain that:

$$\begin{aligned}\dot{\mathbf{z}} &= \mathrm{A_{OI}}\mathbf{z} + \mathrm{b_I} \cdot \beta^{-1}\left(\mathbf{z}\right) \cdot \left[\beta\left(\mathbf{z}\right) \cdot v + \alpha\left(\mathbf{z}\right) - \left(\alpha\left(\mathbf{z}\right) - \beta\left(\mathbf{z}\right) \cdot \left[\underline{\mathrm{a}} \cdot \mathbf{z}\right]\right)\right] \\ &= \mathrm{A_{OI}}\mathbf{z} + \mathrm{b_I}\left[v + \underline{\mathrm{a}} \cdot \mathbf{z}\right].\end{aligned} \tag{2.17}$$

If we derive the scalar equations corresponding to the above, we obtain the following:

$$\mathbf{z} \triangleq \begin{bmatrix} z_1 \\ \vdots \\ z_n \end{bmatrix} \Rightarrow \begin{cases} \dot{z}_1 = z_1^{(1)} = z_2 \\ \dot{z}_2 = z_1^{(2)} = z_3 \\ \vdots \\ \dot{z}_{n-1} = z_1^{(n-1)} = z_n \\ \dot{z}_n = z_1^{(n)} = v + \sum\limits_{\iota=1}^{n} a_{\iota-1} z_\iota = v + \sum\limits_{\iota=0}^{n-1} a_\iota z_1^{(\iota)}. \end{cases} \tag{2.18}$$

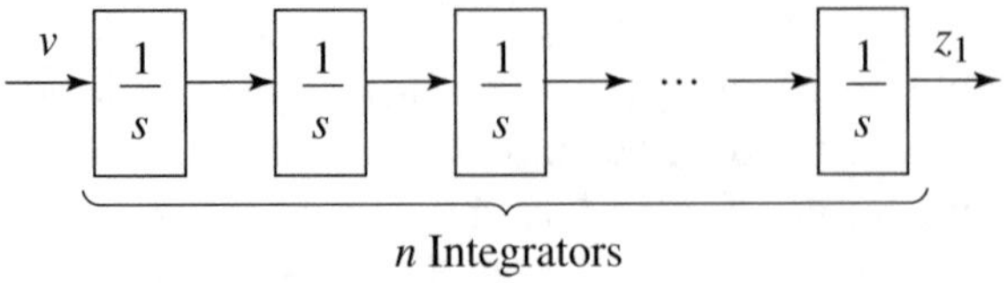

Figure 2.1: Input-state linearization.

The latter in the above becomes effectively a linear scalar ordinary differential equation in z_1. Without an outer loop, i.e., with $\underline{a}$ set to zero, as shown in Fig. 2.1 one obtains:

$$\dot{\mathbf{z}} = \mathrm{A}_{\mathrm{OI}}\,\mathbf{z} + \mathrm{b}_{\mathrm{I}} v \Leftrightarrow \begin{cases} \dot{z}_1 = z_1^{(1)} = z_2 \\ \dot{z}_2 = z_1^{(2)} = z_3 \\ \vdots \\ \dot{z}_{n-1} = z_1^{(n-1)} = z_n \\ \dot{z}_n = z_1^{(n)} = v. \end{cases} \tag{2.19}$$

To derive the transformation $\boldsymbol{T}$ it suffices needs to consider the latter without an outer loop present. Indeed, if $\boldsymbol{T}$ is further broken down to scalar components in a similar pattern as for $\mathbf{z}$ the following can be obtained:

$$\mathbf{z} = \boldsymbol{T}(\mathbf{x}) \triangleq \begin{bmatrix} T_1(\mathbf{x}) \\ \vdots \\ T_n(\mathbf{x}) \end{bmatrix} \Rightarrow \frac{\partial \boldsymbol{T}}{\partial \mathbf{x}} \triangleq \partial_{\mathbf{x}} \boldsymbol{T} = \begin{bmatrix} \frac{\partial T_1}{\partial \mathbf{x}} \\ \vdots \\ \frac{\partial T_n}{\partial \mathbf{x}} \end{bmatrix}. \tag{2.20}$$

So the first of Equations (2.11) becomes:

$$\frac{\partial \boldsymbol{T}}{\partial \mathbf{x}} \cdot \mathbf{f}(\mathbf{x}) = \mathbf{A} \cdot \boldsymbol{T}(\mathbf{x}) - \mathbf{B} \cdot \boldsymbol{\beta}^{-1}(\boldsymbol{T}(\mathbf{x})) \cdot \boldsymbol{\alpha}(\boldsymbol{T}(\mathbf{x}))$$

$$\Downarrow$$

$$\begin{bmatrix} \frac{\partial T_1}{\partial \mathbf{x}} \\ \vdots \\ \frac{\partial T_n}{\partial \mathbf{x}} \end{bmatrix} \cdot \mathbf{f}(\mathbf{x}) = \mathrm{A}_{\mathrm{OI}} \cdot \boldsymbol{T}(\mathbf{x}) - \mathrm{b}_{\mathrm{I}} \cdot \beta^{-1}(\boldsymbol{T}(\mathbf{x})) \cdot \alpha(\boldsymbol{T}(\mathbf{x}))$$

$$\Downarrow$$

$$\begin{bmatrix} \frac{\partial T_1}{\partial \mathbf{x}} \cdot \mathbf{f}(\mathbf{x}) \\ \vdots \\ \frac{\partial T_n}{\partial \mathbf{x}} \cdot \mathbf{f}(\mathbf{x}) \end{bmatrix} = \begin{bmatrix} 0 & 1 & \dots & 0 \\ 0 & \vdots & \ddots & \vdots \\ \vdots & 0 & \dots & 1 \\ 0 & 0 & \dots & 0 \end{bmatrix} \begin{bmatrix} T_1(\mathbf{x}) \\ \vdots \\ T_n(\mathbf{x}) \end{bmatrix} - \begin{bmatrix} 0 \\ 0 \\ \vdots \\ 1 \end{bmatrix} \frac{\alpha(\mathbf{z})}{\beta(\mathbf{z})}.$$

So, eventually one derives the following conditions:

$$\begin{aligned} &\frac{\partial T_1}{\partial \mathbf{x}} \cdot \mathbf{f}(\mathbf{x}) = T_2(\mathbf{x}), \quad \frac{\partial T_2}{\partial \mathbf{x}} \cdot \mathbf{f}(\mathbf{x}) = T_3(\mathbf{x}), \dots, \frac{\partial T_{n-1}}{\partial \mathbf{x}} \cdot \mathbf{f}(\mathbf{x}) = T_n(\mathbf{x}); \\ &\frac{\partial T_n}{\partial \mathbf{x}} \cdot \mathbf{f}(\mathbf{x}) = -\frac{\alpha(\mathbf{z})}{\beta(\mathbf{z})}. \end{aligned} \tag{2.21}$$

Then, the second of Equations (2.11) becomes:

$$\frac{\partial \boldsymbol{T}}{\partial \mathbf{x}} \cdot \mathbf{g}(\mathbf{x}) = \mathbf{B} \cdot \boldsymbol{\beta}^{-1}(\boldsymbol{T}(\mathbf{x}))$$

$$\Downarrow$$

$$\begin{bmatrix} \frac{\partial T_1}{\partial \mathbf{x}} \\ \vdots \\ \frac{\partial T_n}{\partial \mathbf{x}} \end{bmatrix} \cdot \mathbf{g}(\mathbf{x}) = \mathrm{b}_{\mathrm{I}} \cdot \beta^{-1}(\boldsymbol{T}(\mathbf{x}))$$

$$\Downarrow$$

$$\begin{bmatrix} \frac{\partial T_1}{\partial \mathbf{x}} \cdot \mathbf{g}(\mathbf{x}) \\ \vdots \\ \frac{\partial T_n}{\partial \mathbf{x}} \cdot \mathbf{g}(\mathbf{x}) \end{bmatrix} = \begin{bmatrix} 0 \\ 0 \\ \vdots \\ \beta^{-1}(\boldsymbol{T}(\mathbf{x})) \end{bmatrix}.$$

So eventually one derives the following conditions:

$$\frac{\partial T_1}{\partial \mathbf{x}} \cdot \mathbf{g}(\mathbf{x}) = 0, \quad \frac{\partial T_2}{\partial \mathbf{x}} \cdot \mathbf{g}(\mathbf{x}) = 0, \dots, \frac{\partial T_{n-1}}{\partial \mathbf{x}} \cdot \mathbf{g}(\mathbf{x}) = 0; \quad \frac{\partial T_n}{\partial \mathbf{x}} \cdot \mathbf{g}(\mathbf{x}) = \frac{1}{\beta(\mathbf{z})}. \tag{2.22}$$

In conclusion, to summarize for a single-input system, the equations that need to hold for it to be input-state linearizable are:

$$\begin{aligned} &\frac{\partial T_\iota}{\partial \mathbf{x}} \cdot \mathbf{f}(\mathbf{x}) = T_{\iota+1}; \quad \frac{\partial T_\iota}{\partial \mathbf{x}} \cdot \mathbf{g}(\mathbf{x}) = 0, \quad \iota = 1, 2, \dots, n-2, n-1 \\ &\frac{\partial T_n}{\partial \mathbf{x}} \cdot \mathbf{f}(\mathbf{x}) = -\frac{\alpha}{\beta}; \quad \frac{\partial T_n}{\partial \mathbf{x}} \cdot \mathbf{g}(\mathbf{x}) = \frac{1}{\beta} \neq 0. \end{aligned} \tag{2.23}$$

If we can find a scalar transformation multivariable function $T_1(\mathbf{x})$ that satisfies all the requirements above then functions α and β can be calculated as follows:

$$\beta^{-1} = \frac{\partial T_n}{\partial \mathbf{x}} \cdot \mathbf{g}(\mathbf{x}) = \partial_{\mathbf{x}} T_n \cdot \mathbf{g}(\mathbf{x}) \Leftrightarrow \beta = \frac{1}{\partial_{\mathbf{x}} T_n \cdot \mathbf{g}(\mathbf{x})}; \quad \alpha = -\frac{\partial_{\mathbf{x}} T_n \cdot \mathbf{f}(\mathbf{x})}{\partial_{\mathbf{x}} T_n \cdot \mathbf{g}(\mathbf{x})}. \tag{2.24}$$

2.4 THEORY OF INPUT-STATE LINEARIZATION

A general theorem has been developed pertaining to input-state linearization. The preliminary mathematical concepts include the Lie bracket along with others from manifold theory [9]. The Lie bracket (or Lie derivative) is an anticommutative, bilinear, first-order differential operator on vector fields. It may be defined either in terms of local coordinates or in a global, coordinate-free fashion [8]. Though both definitions are prevalent, it is perhaps more straightforward in our case to formulate the Lie Bracket in local coordinates at a point x:

$$L_{\mathrm{f}}\mathrm{g} = [\mathrm{f}, \mathrm{g}] = \mathrm{ad}_{\mathrm{f}}\mathrm{g} = \frac{\partial \mathrm{g}}{\partial \mathrm{x}} \cdot \mathrm{f} - \frac{\partial \mathrm{f}}{\partial \mathrm{x}} \cdot \mathrm{g}. \tag{2.25}$$

In the above f, g are vector fields, i.e., vector functions, $\mathrm{f}, \mathrm{g} : \mathbb{R}^n \to \mathbb{R}^n$; f, g are, moreover, smooth vector field, i.e., they both have continuous partial derivatives of any required order. $\frac{\partial \mathrm{f}}{\partial \mathrm{x}}$ and $\frac{\partial \mathrm{g}}{\partial \mathrm{x}}$ are the Jacobian derivatives of vector field f and g, respectively. Symbol ad in the notation above stands for adjoint. Some further properties of Lie brackets are given below:

$$\mathrm{ad}_{\mathrm{f}}^{0}\,\mathrm{g} = \mathrm{g}; \quad \mathrm{ad}_{\mathrm{f}}^{i+1}\,\mathrm{g} = [\mathrm{f}, \mathrm{ad}_{\mathrm{f}}^{i}\,\mathrm{g}], \quad i = 1, 2, 3, \ldots. \tag{2.26}$$

The Lie derivative of a smooth scalar field (or function, i.e., $\mathbb{R}^n \to \mathbb{R}$) h with respect to a smooth vector field fcan also be defined as follows:

$$L_{\mathrm{f}}h = \frac{\partial h}{\partial \mathrm{x}} \cdot \mathrm{f} = \mathrm{f} \cdot \mathrm{grad}\, h. \tag{2.27}$$

The Lie derivative of a scalar field is effectively a directional derivative and also a scalar. In the above grad or $\frac{\partial h}{\partial \mathrm{x}}$ is the gradient of h. Notice that if g is another smooth vector field and $i = 1, 2, 3, \ldots$:

$$L_{\mathrm{g}}L_{\mathrm{f}}h = \mathrm{grad}\,(L_{\mathrm{f}}h) \cdot \mathrm{g}; \quad L_{\mathrm{f}}^{0}h = h; \quad L_{\mathrm{f}}^{i+1}h = L_{\mathrm{f}}(L_{\mathrm{f}}^{i}h) = \mathrm{f} \cdot \mathrm{grad}\,(L_{\mathrm{f}}^{i}h). \tag{2.28}$$

In contrast to the Lie derivative of a scalar field that is also a scalar, the Lie bracket [f, g] is a vector field. This is clearly seen by the fact that Jacobian derivatives are square matrices of size n equal to the dimension of f and g.

So now assume that vector field [f, g] lies in the plane formed by vectors f and g. Then the set of vector fields {f, g} is called *involutive* [8, 9]. In other words, set {f, g} is involutive if the Lie bracket of f and g, [f, g], can be expressed as a linear combination of f and g.

We use these concepts now to state a general result about the conditions for the linearizability of a single-input linear in control (affine) system of the form encountered before [9]:

$$\dot{\mathbf{x}} = \mathbf{f}(\mathbf{x}) + \mathbf{g}(\mathbf{x}) \cdot \mathbf{u}. \tag{2.29}$$

The system above is input-state linearizable if and only if there exists a region Ω where:

- vector fields $\{\mathbf{g},\ \mathrm{ad}_{\mathrm{f}}\,\mathbf{g}, \ldots, \mathrm{ad}_{\mathrm{f}}^{(n-1)}\mathbf{g}\}$ are linearly independent in Ω and
- the set $\{\mathbf{g},\ \mathrm{ad}_{\mathrm{f}}\,\mathbf{g}, \ldots, \mathrm{ad}_{\mathrm{f}}^{(n-2)}\mathbf{g}\}$ is involutive in Ω.

The first condition can be interpreted as a controllability condition. In the case of a linear system the condition becomes equivalent to the requirement per which the controllability matrix $[\mathrm{B}\ \mathrm{AB} \ldots\ \mathrm{A}^{n-1}\ \mathrm{B}]$ is full rank.

The second condition is also always satisfied by linear systems since the vector fields are constant. But for a generic nonlinear system, as in Equation (2.29), this condition is not always satisfied. On the other hand, it is quite necessary so that according to Frobenius theorem [8, 9] a transformation $T_1(\mathbf{x})$ as required in the previous section exists.

2.5 THEORY OF INPUT-OUTPUT LINEARIZATION

We will now consider a Single-Input-Single-Output (SISO) system of the following form:

$$\begin{aligned} \dot{\mathbf{x}} &= \mathbf{f}(\mathbf{x}) + \mathbf{g}(\mathbf{x}) \cdot u \\ y &= h(\mathbf{x}). \end{aligned} \tag{2.30}$$

In the above, **f**, **g**, and h are sufficiently smooth in a domain $\boldsymbol{D} \subset \mathbb{R}^n$. Mappings (i.e., functions) **f** and **g** are vector fields as the term was introduced in the previous section while function h is a scalar field. We will now proceed to derive conditions that allow us to transform the system equations so that the input-output mapping is linear.

Based on the theory of the previous section and using the Lie derivative of a scalar field with respect to a vector field:

$$\dot{y} \equiv y^{(1)} \equiv \frac{d}{dt} y = \frac{\partial h}{\partial \mathbf{x}} \cdot \dot{\mathbf{x}} = \frac{\partial h}{\partial \mathbf{x}} \cdot [\mathbf{f}(\mathbf{x}) + \mathbf{g}(\mathbf{x}) \cdot u] = L_{\mathrm{f}} h(\mathbf{x}) + L_{\mathrm{g}} h(\mathbf{x}) \cdot u.$$

If $L_{\mathrm{g}} h$ is zero then $y^{(1)} = L_{\mathrm{f}} h$ and evidently independent of input u. We then continue to obtain the second time derivative of the output y denoted by $y^{(2)}$. Thus, we obtain:

$$\ddot{y} \equiv y^{(2)} = \frac{\partial\,(L_{\mathrm{f}} h)}{\partial \mathbf{x}} \cdot \dot{\mathbf{x}} = \frac{\partial\,(L_{\mathrm{f}} h)}{\partial \mathbf{x}} \cdot [\mathbf{f}(\mathbf{x}) + \mathbf{g}(\mathbf{x}) \cdot u] = L_{\mathrm{f}}^2 h(\mathbf{x}) + L_{\mathrm{g}} L_{\mathrm{f}} h(\mathbf{x}) \cdot u.$$

In the above some of the properties of Lie derivatives have been employed. If now $L_{\mathrm{g}} L_{\mathrm{f}} h$ is again zero then $y^{(2)} = L_{\mathrm{f}}^2 h$ and also independent of u. Repeating this process, we can verify that if

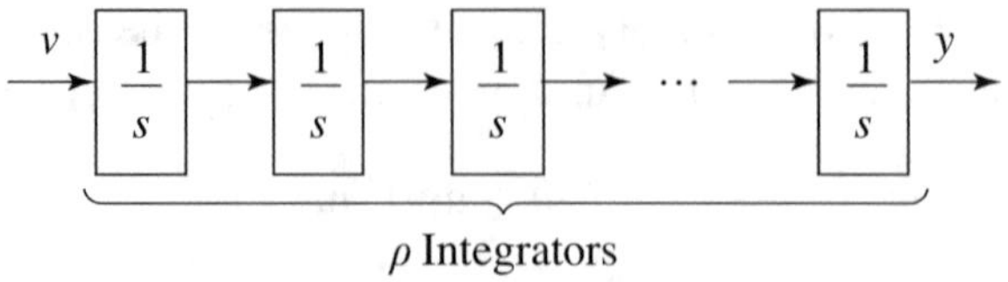

Figure 2.2: Input-output linearization.

$h(\mathbf{x})$ satisfies the following:

$$\begin{aligned} &L_{\mathrm{g}} L_{\mathrm{f}}^{\iota} h(\mathbf{x}) = 0, \quad \iota = 1, 2, \ldots, \rho - 2 \\ &L_{\mathrm{g}} L_{\mathrm{f}}^{\rho-1} h(\mathbf{x}) \neq 0. \end{aligned} \tag{2.31}$$

Then u does not appear in the equations for $y, y^{(1)}, \ldots, y^{(\rho-1)}$ but it appears in the equation for $y^{(\rho)}$ with a nonzero coefficient, as follows:

$$y^{(\rho)} = L_{\mathrm{f}}^{\rho} h(\mathbf{x}) + L_{\mathrm{g}} L_{\mathrm{f}}^{\rho-1} h(\mathbf{x}) \cdot u. \tag{2.32}$$

The equation above shows clearly that the system is *input-output linearizable*. Indeed, the following nonlinear full state feedback control law reduces the input-output map to a chain of ρ in number integrators; see Fig. 2.2.

$$u = \frac{v - L_{\mathrm{f}}^{\rho} h(\mathbf{x})}{L_{\mathrm{g}} L_{\mathrm{f}}^{\rho-1} h(\mathbf{x})} \Rightarrow y^{(\rho)} = v. \tag{2.33}$$

Effectively we have defined a new input v such that y is related to it by linear dynamics in (at least) some subset $\boldsymbol{D} \subset \mathbb{R}^n$ of the state space. We do this so that in an outer loop we can select a control law v with linear methods. Then v is transformed by Equation (2.33) to the original control u. Integer ρ is called the *relative degree* of the system at hand [9]. For linear systems, relative degree ρ is the difference between the number of poles and zeros in the transfer function of the system.

It is possible, however, that at one or more isolated points, $\mathbf{x}_0 \in \boldsymbol{D}$, $L_{\mathrm{g}} L_{\mathrm{f}}^{\rho-1} h$ is zero while nonzero in every other point in $\boldsymbol{D}$. This makes the relative degree *undefined*. This can be of importance if one of those $\mathbf{x}_0 \in \boldsymbol{D}$ points are of interest, e.g., if it is the desired point of regulation or if system state trajectories cannot be guaranteed to approach $\mathbf{x}_0$. If input-output linearization design is conducted ignoring this fact, unwanted and erroneous consequences may occur.

When the relative degree of a system is equal to its order, i.e., when $\rho = n$, then input-output linearization is essentially equivalent to input-state linearization [9]. When, however, $\rho < n$ then there are *internal dynamics* not expressed in the input-output relationship. It is interesting to see some more facts about internal dynamics and the relative degree, especially in the case of linear systems. We then proceed to generalize in the case of nonlinear systems.

2.6 INTERNAL AND ZERO DYNAMICS

Consider a SISO linear, time-invariant (LTI) system with transfer function as follows:

$$H(s) = \frac{Y(s)}{U(s)} = \frac{b_m \cdot s^m + b_{m-1} \cdot s^{m-1} + \cdots + b_1 \cdot s + b_0}{s^n + a_{n-1} \cdot s^{n-1} + \cdots + a_1 \cdot s + a_0}, \quad s \in \mathbb{C}. \tag{2.34}$$

Assuming the system above is strictly proper, i.e., $m < n$ and of course $b_m \neq 0$, a state-space realization of the system is the *controllable companion canonical form* given below [10–16]:

$$\begin{aligned} \dot{\mathrm{x}} &= \mathrm{A} \cdot \mathrm{x} + \mathrm{B} \cdot u \\ y &= \mathrm{C} \cdot \mathrm{x} + Du. \end{aligned} \tag{2.35}$$

In the above, A is an $n \times n$ matrix, B an $n \times 1$ column matrix, C a $1 \times n$ row matrix, and D a scalar since the system is assumed to be LTI and to have order of n as demonstrated in its transfer function. Finally, the state equations above are expanded below to better reveal the structure of the SISO-LTI system considered:

$$\begin{aligned} \begin{bmatrix} \dot{x}_1 \\ \dot{x}_2 \\ \dot{x}_3 \\ \vdots \\ \dot{x}_n \end{bmatrix} &= \begin{bmatrix} 0 & 1 & 0 & \cdots & 0 \\ 0 & 0 & 1 & \cdots & 0 \\ 0 & 0 & 0 & \cdots & 0 \\ \vdots & \vdots & \vdots & \ddots & \vdots \\ -a_0 & -a_1 & -a_2 & \cdots & -a_{n-1} \end{bmatrix} \cdot \begin{bmatrix} x_1 \\ \vdots \\ x_m \\ \vdots \\ x_n \end{bmatrix} + \begin{bmatrix} 0 \\ 0 \\ 0 \\ \vdots \\ 1 \end{bmatrix} \cdot u \\ y &= \begin{bmatrix} b_0 & \ldots & b_m & \ldots & 0 \end{bmatrix} \cdot \begin{bmatrix} x_1 \\ \vdots \\ x_m \\ \vdots \\ x_n \end{bmatrix} \end{aligned} \tag{2.36}$$

The expanded form reveals that $D = 0$ since the system is strictly proper. Furthermore, it is evidently (state) controllable [10–16], i.e., the controllability matrix $[\mathrm{B}\ \mathrm{AB} \ldots \mathrm{A}^{n-1}\mathrm{B}]$ is full rank.

Now clearly the LTI-SISO system is a special case of the nonlinear system in Equation (2.30) with the following analogies:

$$\mathbf{f(x)} = \mathrm{A} \cdot \mathrm{x}, \quad \mathbf{g(x)} = \mathrm{B}; \quad h(\mathbf{x}) = \mathrm{C} \cdot \mathrm{x}.$$

Let us now apply the successive differentiation procedure as previously:

$$y^{(1)} = \frac{\partial h}{\partial \mathbf{x}} \cdot \dot{\mathbf{x}} = \mathrm{CAx} + \mathrm{CB}\, u.$$

In the case $m = n - 1$, $\mathrm{CB} = b_{n-1} = b_m$ and the system has relative degree $\rho = 1$. Otherwise, $\mathrm{CB} = 0$ and the process needs to be continued farther to obtain $y^{(2)}$. Also, note that CA is a row matrix where the elements of C have just been shifted one position to the right; in general, CA^{ι} is also a row matrix where the elements of C are just shifted one position to the right if $\iota = 0, \dots, n - m - 1$. In conclusion, $\mathrm{CA}^{\iota-1}\,\mathrm{B} = 0$ for $\iota = 0, \dots, n - m - 1$ and $\mathrm{CA}^{n-m-1}\,\mathrm{B} = b_m \neq 0$. In effect, u appears for the first time in the equation for $y^{(n-m)}$, as follows:

$$y^{(n-m)} = \mathrm{CA}^{n-m}\,\mathrm{x} + \mathrm{CA}^{n-m-1}\,\mathrm{B}\,u. \tag{2.37}$$

Thus, the relative degree of the LTI-SISO system at hand is $n - m$. Note that this is equal to the difference of the degree of the denominator polynomial (also known as characteristic polynomial of the system [10–16]) minus that of the numerator polynomial of the system transfer function $H(s)$. One can see that the roots of the numerator polynomial, also referred to as the zeros of the transfer function in the LTI-SISO case [10–16], correspond to eigenvalues of internal dynamics if any. Indeed, internal dynamics is this part of the system dynamics that arise even when the output is constrained to be zero over all times; this is why another term used in literature for this part of dynamics is *zero dynamics* [17–19]. In the SISO-LTI case specifically the following partition and transformation of the state vector can be applied:

$$\chi = \begin{bmatrix} \mathrm{z} \\ \hline \boldsymbol{v} \end{bmatrix} \begin{bmatrix} z_1 \\ \vdots \\ z_m \\ \hline v_1 \\ \vdots \\ v_{n-m} \end{bmatrix} = \begin{bmatrix} x_1 \\ \vdots \\ x_m \\ \hline y \\ \vdots \\ y^{(\rho-1)} \end{bmatrix} = \begin{bmatrix} x_1 \\ \vdots \\ x_m \\ \hline b_0 x_1 + b_1 x_2 + \dots + b_m x_{m+1} \\ \vdots \\ b_0 x_{n-m} + b_1 x_{\rho+1} + \dots + b_m x_n \end{bmatrix}. \tag{2.38}$$

Then for component vector υ the following holds:

$$\frac{d\boldsymbol{v}}{dt} = \begin{bmatrix} \dot{y} \\ \vdots \\ y^{(n-m)} \end{bmatrix} = \begin{bmatrix} v_2 \\ \vdots \\ \dot{v}_{n-m} \end{bmatrix} = \begin{bmatrix} \mathrm{CA}\,\mathrm{x} \\ \vdots \\ \mathrm{CA}^{n-m}\,\mathrm{x} + \mathrm{CA}^{\rho-1}\,\mathrm{B}\,u \end{bmatrix}. \tag{2.39}$$

The last row in the above is broken down further as follows:

$$\begin{aligned} \dot{v}_{n-m} &= y^{(n-m)} = b_0 \dot{x}_{n-m} + b_1 \dot{x}_{\rho+1} + \dots + b_m \dot{x}_n \\ &= b_0 x_{\rho+1} + b_1 x_{\rho+2} + \dots + b_m \left(u - a_0 x_1 + a_1 x_2 + \dots + a_n x_n\right) \\ &= \underline{\mathrm{m}}_{\mathrm{z}}\,\mathrm{z} + \underline{\mathrm{m}}_{v}\,\boldsymbol{v} + b_m u. \end{aligned} \tag{2.40}$$

Row vectors $\underline{\mathrm{m}}_{\mathrm{z}}$ and $\underline{\mathrm{m}}_{\boldsymbol{v}}$ are straightforward to obtain from Equations (2.36) and (2.38). The most important consequence of Equation (2.40), however, is that the following state-feedback control law can be employed to implement the constraint $y^{(n-m)}$ is zero constantly over time:

$$u = -\frac{1}{b_m}\left(\underline{\mathrm{m}}_z \mathrm{z} + \underline{\mathrm{m}}_v \boldsymbol{v}\right), \quad b_m \neq 0, \quad \rho = n - m > 0. \tag{2.41}$$

Now because of Equation (2.39) and assuming that initial conditions for y and its derivatives up to order $(\rho - 1)$ are all zero, too, the constraint that y is zero over all time is guaranteed. We now turn our attention to the first part of state vector χ in Equation (2.38); this part defines the zero dynamics. Indeed:

$$\frac{d\mathrm{z}}{dt} = \begin{bmatrix} \dot{x}_1 \\ \vdots \\ \dot{x}_m \end{bmatrix} = \begin{bmatrix} x_2 \\ \vdots \\ x_{m+1} \end{bmatrix} = \mathrm{M}\mathrm{z} + \mathrm{M}_{\mathrm{O}}\boldsymbol{v}$$

$$\Updownarrow$$

$$\frac{d\mathrm{z}}{dt} = \begin{bmatrix} 0 & 1 & 0 & \dots & 0 \\ 0 & 0 & 1 & \dots & 0 \\ 0 & 0 & 0 & \dots & 0 \\ \vdots & \vdots & \vdots & \ddots & \vdots \\ -\frac{b_0}{b_m} & -\frac{b_1}{b_m} & -\frac{b_2}{b_m} & \dots & -\frac{b_{m-1}}{b_m} \end{bmatrix} \mathrm{z} + \begin{bmatrix} 0 & 0 & 0 & \dots & 0 \\ 0 & 0 & 0 & \dots & 0 \\ 0 & 0 & 0 & \dots & 0 \\ \vdots & \vdots & \vdots & \ddots & \vdots \\ \frac{1}{b_m} & 0 & 0 & \dots & 0 \end{bmatrix} \boldsymbol{v}\,. \tag{2.42}$$

Given now that control input u will be set as in Equation (2.41) to guarantee the constraint that y is zero, we obtain finally that:

$$\frac{d\mathrm{z}}{dt} = \begin{bmatrix} 0 & 1 & 0 & \dots & 0 \\ 0 & 0 & 1 & \dots & 0 \\ 0 & 0 & 0 & \dots & 0 \\ \vdots & \vdots & \vdots & \ddots & \vdots \\ -\frac{b_0}{b_m} & -\frac{b_1}{b_m} & -\frac{b_2}{b_m} & \dots & -\frac{b_{m-1}}{b_m} \end{bmatrix} \mathrm{z}. \tag{2.43}$$

This concludes the demonstration for LTI-SISO system in the controllable companion canonical form. Indeed, as shown above, even though the output may be identically zero over all times, there is an internal part to the dynamics (thereof the term *internal dynamics*) that may be active. Notice that stability and all other important properties for zero dynamics of an LTI-SISO are determined by the zeros of the transfer function. Indeed, matrix M is in the controllable companion canonical form and, in effect, the eigenvalues of zero dynamics are the roots of the following polynomial:

$$q_o(s) = b_m s^m + b_{m-1} s^{m-1} + \cdots + b_1 s + b_0. \tag{2.44}$$

This is why *non-minimum phase systems*, i.e., systems with unstable zeros [10–16], are considered rather problematic in linear control system theory; the zero dynamics can give rise to instabilities even if the output is identically zero. In the linear case, non-minimum phase systems have two salient features: (a) they have at least one zero in their transfer function; and (b) they exhibit a larger phase variation with minimum-phase system counterparts, identified by the fact that the system transfer function magnitudes coincide at all frequencies (thereof the term minimum phase). Zeros in multivariable LTI systems require a more involved definition; see *transmission zeros* [10–16].

The concept of internal (or zero) dynamics carries over nicely to the more general framework of nonlinear systems, as well [17–19]. In this case, an affine SISO system of the following form will be considered:

$$\begin{aligned} \dot{\mathbf{x}} &= \mathbf{f}(\mathbf{x}) + \mathbf{g}(\mathbf{x}) \cdot u \\ y &= h(\mathbf{x}). \end{aligned} \tag{2.45}$$

In the nonlinear case, the partition as well as the transformation of the state vector to the *normal form* [10–19] is done on the basis of relative degree ρ. Specifically, the following transformation will be considered:

$$\chi = \begin{bmatrix} \mathrm{z} \\ \hline\hline \boldsymbol{v} \end{bmatrix} = \begin{bmatrix} z_1 \\ \vdots \\ z_{n-\rho} \\ \hline\hline v_1 \\ \vdots \\ v_\rho \end{bmatrix} = \begin{bmatrix} \varphi_1(\mathbf{x}) \\ \vdots \\ \varphi_{n-\rho}(\mathbf{x}) \\ \hline\hline h(\mathbf{x}) \\ \vdots \\ L_{\mathbf{f}}^{(\rho-1)} h(\mathbf{x}) \end{bmatrix}. \tag{2.46}$$

For vector z in the above the following must hold at least in a domain $\boldsymbol{D}$ as discussed earlier:

$$L_{\mathbf{g}}\boldsymbol{\varphi}_\iota(\mathbf{x}) = \frac{\partial \boldsymbol{\varphi}_\iota}{\partial \mathbf{x}} \cdot \mathbf{g} = 0, \quad 1 \le \iota \le n - \rho. \tag{2.47}$$

Due to the above condition, input u vanishes in the dynamics of z:

$$\frac{d\mathrm{z}}{dt} = \frac{\partial \mathrm{z}}{\partial \mathbf{x}} \cdot [\mathbf{f}(\mathbf{x}) + \mathbf{g}(\mathbf{x}) \cdot u] = \frac{\partial \mathrm{z}}{\partial \mathbf{x}} \cdot \mathbf{f}(\mathbf{x}) + \underbrace{\frac{\partial \mathrm{z}}{\partial \mathbf{x}} \cdot \mathbf{g}(\mathbf{x})}_{0} \cdot u = \frac{\partial \mathrm{z}}{\partial \mathbf{x}} \cdot \mathbf{f}(\mathbf{x}). \tag{2.48}$$

Therefore, z represents the internal (or zero) dynamics of the system, Furthermore, the following definition can be introduced:

$$\mathrm{f}_0(\mathrm{z}, \boldsymbol{v}) \triangleq \frac{\partial \mathrm{z}}{\partial \mathbf{x}} \cdot \mathbf{f}(\mathbf{x}). \tag{2.49}$$

For vector $\boldsymbol{v}$ the definition of the *relative degree* for the system at hand needs to be recalled [9]:

$$\begin{aligned} &L_{\mathrm{g}} L_{\mathrm{f}}^{\iota} h(\mathbf{x}) = 0, \quad \iota = 1, 2, \ldots, \rho - 2 \\ &L_{\mathrm{g}} L_{\mathrm{f}}^{\rho-1} h(\mathbf{x}) \neq 0. \end{aligned} \tag{2.50}$$

Then u does not appear in the equations for $y, y^{(1)}, \ldots, y^{(\rho-1)}$ but it appears in the equation for $y^{(\rho)}$ with a nonzero coefficient, as follows:

$$y^{(\rho)} = L_{\mathrm{f}}^{\rho} h(\mathbf{x}) + L_{\mathrm{g}} L_{\mathrm{f}}^{\rho-1} h(\mathbf{x}) \cdot u. \tag{2.51}$$

So by following the procedure already outlined one can linearize the system, in the input-output sense, using the following full state feedback control law and the "synthetic" input v:

$$u = \frac{v - L_{\mathrm{f}}^{\rho} h(\mathbf{x})}{L_{\mathrm{g}} L_{\mathrm{f}}^{\rho-1} h(\mathbf{x})} \Rightarrow y^{(\rho)} = v. \tag{2.52}$$

So now the system equations assume the following form:

$$\begin{aligned} \frac{d\mathrm{z}}{dt} &= \mathrm{f}_0(\mathrm{z}, \boldsymbol{v}) \\ \frac{d\boldsymbol{v}}{dt} &= \underbrace{\begin{bmatrix} 0 & 1 & \ldots & 0 \\ 0 & \vdots & \ddots & \vdots \\ \vdots & 0 & \ldots & 1 \\ 0 & 0 & \ldots & 0 \end{bmatrix}}_{\rho \times \rho} \cdot \boldsymbol{v} + \underbrace{\begin{bmatrix} 0 \\ 0 \\ \vdots \\ 1 \end{bmatrix}}_{\rho \times 1} \cdot \underbrace{\beta_o(\mathrm{z}, \boldsymbol{v}) \cdot (u - \alpha_o(\mathrm{z}, \boldsymbol{v}))}_{=v} \\ y &= \underbrace{\begin{bmatrix} 1 & 0 & \ldots & 0 \end{bmatrix}}_{1 \times \rho} \cdot \boldsymbol{v} = v_1. \end{aligned} \tag{2.53}$$

In the above,

$$\beta_o(\mathrm{z}, \boldsymbol{v}) = L_{\mathrm{g}} L_{\mathrm{f}}^{\rho-1} h(\mathbf{x}), \quad \alpha_o(\mathrm{z}, \boldsymbol{v}) = -\frac{L_{\mathrm{f}}^{\rho} h(\mathbf{x})}{L_{\mathrm{g}} L_{\mathrm{f}}^{\rho-1} h(\mathbf{x})}. \tag{2.54}$$

One can easily verify that the dynamics for $\boldsymbol{v}$ is a canonical form representation of a chain of ρ integrators. Notice that functions α_o and β_o are expressed in coordinates of transformed state χ, or equivalently, vectors z and $\boldsymbol{v}$. This is possible because they are independent of the choice of the φ's that make up z and are uniquely determined in terms of **f**, **g**, and h. This can be achieved by substituting for **x** on the right-hand side of Equations (2.54), the corresponding state vector χ as identified through the transform introduced in Equation (2.46).

The three new equations of the system as in (2.53) are said to be in the *normal form* [9, 17–19]. In this form the system is decomposed to an external part $\boldsymbol{v}$ and an internal part z. The external part can be linearized per the synthetic input v using the state feedback control law in

Equation (2.52). However, note that the same control law renders the internal part *unobservable* [9–19]!

Let us now consider the internal dynamics of the system at hand:

$$\frac{dz}{dt} = f_0 (z, \boldsymbol{v}) . \tag{2.55}$$

As in the linear case, if $\boldsymbol{v}$ is set to zero the zero dynamics of the system are obtained:

$$\frac{dz}{dt} = f_0 (z, \mathbf{0}) . \tag{2.56}$$

If the zero dynamics of the system above are (globally) asymptotically stable then the nonlinear system is called *minimum phase.*

It can be proven [9, 17–19] that the origin $\mathbf{x} = \mathbf{0}$ is an asymptotically stable equilibrium of the linearized system if $z = \mathbf{0}$ is an asymptotically stable equilibrium point of its zero dynamics. In other words, any minimum phase, input-output linearizable system can be stabilized by a state feedback control law. This is, however, only a local result. Indeed, it is not the case, as one might expect, that because the origin $z = \mathbf{0}$ is stable for the zero dynamics, the system can be globally stabilized [9, 17–19]. Global stability can be guaranteed only if the system is input-to-state stable and linearizable.

The stability of internal and zero dynamics is also referred to in literature [9, 17–19] as *internal stability*. Based on the argumentation before, it is imperative to check internal stability of systems with relative degree less than their order, especially when the controllers are designed to regulate the output to zero, e.g., sliding mode control. Furthermore, note that particularly when performing design on the basis of input-output linearization, stability of any residual internal dynamics is always a concern that needs to be verified.

CHAPTER 3

Exact Linearization of Modulated State Systems

3.1 AN ALTERNATIVE PROBLEM FORMULATION

We return now to the system introduced in Chapter 1 that exhibits nonlinear cross-band coupling and state modulation and demodulation:

$$\dot{\mathbf{x}}_1 = \mathbf{A}_{LP} \cdot \mathbf{x}_1 + \boldsymbol{\psi}(\mathbf{y}_2) + \mathbf{d} \tag{3.1}$$

$$\begin{aligned} \dot{\mathbf{x}}_2 &= \mathbf{F}(\mathbf{x}_1) \cdot \mathbf{x}_2 + \mathbf{G}(\mathbf{x}_1) \cdot \mathbf{u} \\ \mathbf{y}_2 &= \mathbf{C}_2 \cdot \mathbf{x}_2. \end{aligned} \tag{3.2}$$

We will perform a first comparison with the feedback linearization general form to check if it can be applied:

$$\begin{aligned} \dot{\mathbf{x}} &= \mathbf{f}(\mathbf{x}) + \mathbf{g}(\mathbf{x}) \cdot \mathbf{u} \\ \mathbf{y} &= \mathbf{h}(\mathbf{x}). \end{aligned} \tag{3.3}$$

The original system is indeed a special case of the one in Equation (2.1) especially when undisturbed ($\mathbf{d} = \mathbf{0}$). Indeed, the dynamic equation can be rewritten as follows:

$$\begin{aligned} \dot{\mathbf{x}} &= \underbrace{\begin{bmatrix} \mathbf{A}_{LP} & \mathbf{0} \\ \mathbf{0} & \mathbf{F}(\mathbf{x}_1) \end{bmatrix} \mathbf{x} + \begin{bmatrix} \boldsymbol{\psi}(\mathbf{C}_2\mathbf{x}_2) \\ \mathbf{0} \end{bmatrix}}_{\mathrm{f(x)}} + \underbrace{\begin{bmatrix} \mathbf{0} \\ \mathbf{G}(\mathbf{x}_1) \end{bmatrix}}_{\mathrm{g(x)}} \mathbf{u}; \quad \mathbf{x} = \begin{bmatrix} \mathbf{x}_1 \\ \mathbf{x}_2 \end{bmatrix} \\ \mathbf{y} &\equiv \mathbf{y}_2 = \mathbf{C}_2 \cdot \mathbf{x}_2 \equiv h(\mathbf{x}). \end{aligned} \tag{3.4}$$

Also, note that the equivalent LP system introduced earlier is a special case of the one in Equation (2.1) especially when undisturbed ($\mathbf{d} = \mathbf{0}$). Indeed, the dynamic equation can be rewritten as follows:

$$\begin{aligned} \dot{\mathbf{x}} &= \underbrace{\begin{bmatrix} \mathbf{A}_{LP} & \mathbf{0} \\ \mathbf{0} & \mathbf{F}(\mathbf{x}_1) - j\omega_c\mathbf{I} \end{bmatrix} \mathbf{x} + \begin{bmatrix} \boldsymbol{\psi}_{LP}(\mathbf{C}_2\tilde{\mathbf{x}}_2) \\ \mathbf{0} \end{bmatrix}}_{\mathrm{f(x)}} + \underbrace{\begin{bmatrix} \mathbf{0} \\ \mathbf{G}(\mathbf{x}_1) \end{bmatrix}}_{\mathrm{g(x)}} \tilde{\mathbf{u}}; \quad \mathbf{x} = \begin{bmatrix} \mathbf{x}_1 \\ \tilde{\mathbf{x}}_2 \end{bmatrix} \\ \mathbf{y} &\equiv \tilde{\mathbf{y}}_2 = \mathbf{C}_2 \cdot \tilde{\mathbf{x}}_2 \equiv h(\mathbf{x}). \end{aligned} \tag{3.5}$$

So the general procedure presented in the previous chapter could be, at least in principle, attempted. However, an alternative approach will be pursued in the sequel. As can be seen the band-pass (BP) subsystem alone in Equation (3.2) is clearly a special case of the affine nonlinear system structure in Equation (3.3). However, the nonlinear cross-band coupling cannot occur unless the Taylor expansion of $\boldsymbol{\psi}$ hereafter does not contain even power terms; eventually, odd power terms play no role and are filtered out:

$$\boldsymbol{\psi}(\mathbf{y}_2) = \boldsymbol{\psi}(\mathbf{y}_2 = \mathbf{0}) + \left[\sum_{k=1}^{\infty} \boldsymbol{\psi}_i^{(k)} \cdot \mathbf{y}_2^{\otimes k} \right], \quad 1 \leq i \leq n_1. \tag{3.6}$$

This is due to: (a) that transfer function matrix $\mathbf{H}_2(s)$ as follows is BP around a sufficiently high carrier frequency ω_c:

$$\mathbf{H}_2(s) = \mathbf{C}_2 \cdot (s\mathbf{I} - \mathbf{F}_0)^{-1} \cdot \boldsymbol{\Gamma}. \tag{3.7}$$

Also, due to: (b) that transfer function matrix $\mathbf{H}_1(s)$ as follows is LP:

$$\mathbf{H}_1(s) = (s\mathbf{I} - \mathbf{A}_{LP})^{-1}. \tag{3.8}$$

But linearization in any sense, e.g., exact input-state or input-output linearization via feedback or "traditional" approximate linearization around an (equilibrium) point in the state space, would essentially mean that the first power, which is odd would become the predominant feature of the intrinsic system structure. So, in effect, if the LP subsystem is linearizable, too, then cross-band coupling with the BP subsystem through the even powers in the expansion of Equation (3.6) for $\boldsymbol{\psi}$ might not be feasible. This is why the alternative approach is pursued here. For example, appropriate transformations as specified in the previous chapter for input-state or input-output linearization may not exist or, at least, be meaningful to derive.

3.2 INITIAL STRUCTURAL CONSIDERATIONS

We can now look into a simplistic and heuristic approach to BP subsystem exact feedback linearization and see how it fails but also what useful features can be adopted. For simplicity, consider a single-input, two-dimensional case, aka:

$$\mathbf{x}_2 = \begin{bmatrix} x_1 \\ x_2 \end{bmatrix}, \quad \mathbf{F}(\mathbf{x}_1) = \begin{bmatrix} \mathrm{F}_{11}(\mathbf{x}_1) & \mathrm{F}_{12}(\mathbf{x}_1) \\ \mathrm{F}_{21}(\mathbf{x}_1) & \mathrm{F}_{22}(\mathbf{x}_1) \end{bmatrix}, \quad \mathbf{G}(\mathbf{x}_1) = \begin{bmatrix} 0 \\ g(\mathbf{x}_1) \end{bmatrix}, \quad g(\mathbf{x}_1) \neq 0. \tag{3.9}$$

Then, one might be tempted to try the following state feedback control toward exact linearization of the system:

$$\begin{aligned} u &= \frac{1}{g(\mathbf{x}_1)} \left(v - \begin{bmatrix} K_1 & K_2 \end{bmatrix} \mathbf{F}(\mathbf{x}_1)\, \mathbf{x}_2 \right) \\ &= \frac{v - \begin{bmatrix} K_1 \mathrm{F}_{11}(\mathbf{x}_1) + K_2 \mathrm{F}_{21}(\mathbf{x}_1) & K_1 \mathrm{F}_{12}(\mathbf{x}_1) + K_2 \mathrm{F}_{22}(\mathbf{x}_1) \end{bmatrix} \mathbf{x}_2}{g(\mathbf{x}_1)}. \end{aligned} \tag{3.10}$$

In the above, K_1 and K_2 are gains to be determined. However, by substitution in the system's dynamical equation the following is obtained:

$$\dot{\mathbf{x}}_2 = \mathbf{F}(\mathbf{x}_1) \cdot \mathbf{x}_2 + \mathbf{G}(\mathbf{x}_1) \cdot \frac{v - \begin{bmatrix} K_1 & K_2 \end{bmatrix} \mathbf{F}(\mathbf{x}_1)\,\mathbf{x}_2}{g(\mathbf{x}_1)}$$

$$\Updownarrow$$

$$\left\{ \begin{array}{c} \dot{x}_1 = x_1\,\mathrm{F}_{11}(\mathbf{x}_1) + x_2\,\mathrm{F}_{12}(\mathbf{x}_1) \\ \dot{x}_2 = x_1\,\mathrm{F}_{21}(\mathbf{x}_1) + x_2\,\mathrm{F}_{22}(\mathbf{x}_1) + v - (K_1\,\mathrm{F}_{11} + K_2\,\mathrm{F}_{21})\,x_1 - (K_1\,\mathrm{F}_{12} + K_2\,\mathrm{F}_{22})\,x_2 \end{array} \right\}. \tag{3.11}$$

Clearly, setting $K_1 = 0$ and $K_2 = 1$ allows to linearize successfully the second equation of the above for x_2; however, the first equation for x_1 does not include either K_1 or K_2. In effect, it cannot be processed further.

Despite this failure, some elements of this straightforward analysis can be used further. Specifically, assume that $\mathbf{F}$ is of the following form:

$$\mathbf{F}(\mathbf{x}_1) = \begin{bmatrix} \zeta_1 & \zeta_2 \\ \mathrm{F}_{21}(\mathbf{x}_1) & \mathrm{F}_{22}(\mathbf{x}_1) \end{bmatrix}; \quad \zeta_1, \zeta_2 \quad \text{real constants.} \tag{3.12}$$

So effectively $\mathbf{F}$ is partitioned to a real and constant upper block (e.g., ζ_1, ζ_2 can be so that this block corresponds to the Brunovsky canonical form) and a nonlinear lower block depending exclusively on state vector $\mathbf{x}_1$. In this case, after applying the feedback control law of Equation (3.10) with the K's open, we see that the scalar equations for x_1 and x_2 become:

$$\begin{aligned} \dot{x}_1 &= \zeta_1 x_1 + \zeta_2 x_2 \\ \dot{x}_2 &= (\mathrm{F}_{21} - K_1\,\mathrm{F}_{11} - K_2\,\mathrm{F}_{21})\,x_1 + (\mathrm{F}_{22} - K_1\,\mathrm{F}_{12} - K_2\,\mathrm{F}_{22})\,x_2 + v. \end{aligned} \tag{3.13}$$

So now by setting $K_1 = 0$ and $K_2 = 1$ the equations become linear with respect to the synthetic input v as follows:

$$\dot{x}_1 = \zeta_1 x_1 + \zeta_2 x_2, \quad \dot{x}_2 = v. \tag{3.14}$$

A final point is now mentioned concerning the output equation of the BP subsystem

$$\mathbf{y}_2 = \mathbf{C}_2 \cdot \mathbf{x}_2. \tag{3.15}$$

Its linear form may, as seen in the following section, allows dealing with input-state and input-output linearization in a unified fashion.

3.3 A SUFFICIENT BAND-PASS SUBSYSTEM STRUCTURE

We now take advantage of the point made in the end of the previous section after generalizing and properly setting it up in the generic exact linearization framework of the previous chapter.

Specifically, we first require setting the BP subsystem state-space equations into an appropriately modified controllable companion canonical form by taking into advantage the fact that matrix functions $\mathbf{F}$ and $\mathbf{G}$ of the dynamical equation do not depend on $\mathbf{x}_2$ but only on the state vector of the LP subsystem $\mathbf{x}_1$. This form will prove valuable to solve the partial linearization problem:

$$\dot{\mathbf{x}}_2 = \mathbf{F}(\mathbf{x}_1)\cdot\mathbf{x}_2 + \mathbf{G}(\mathbf{x}_1)\cdot\mathbf{u}$$

$$\mathbf{F}(\mathbf{x}_1) = \left[\begin{array}{c}\mathbf{F}_2\\ \hline\hline \mathbf{F}_1(\mathbf{x}_1)\end{array}\right], \quad \mathbf{G}(\mathbf{x}_1) = \left[\begin{array}{c}\mathbf{0}_{\rho_2\times m_2}\\ \hline\hline \mathbf{G}_1(\mathbf{x}_1)\end{array}\right], \quad \rho_2 = n_2 - m_2 \geq 0. \tag{3.16}$$

In the above:

$\rho_2 = n_2 - m_2$, i.e., the excess in number of states vs. inputs of the BP subsystem.

$\mathbf{F}_2$ is a $\rho_2 \times n_2$ constant matrix.

$\mathbf{F}_1$ is a $m_2 \times n_2$ matrix function of LP subsystem state vector $\mathbf{x}_1$.

$\mathbf{0}$ is the zero matrix or vector of specified dimensions.

$\mathbf{G}_1$ is a $m_2 \times m_2$ matrix function of LP subsystem state vector $\mathbf{x}_1$; $\mathbf{G}_1$ must be invertible in at least a domain $\mathbf{D}_1 \subset \mathbb{R}^\kappa$ where $\kappa = n_1$ is the dimension of the LP subsystem state vector $\mathbf{x}_1$.

Notice the similarity of the structure considered to the controllable companion canonical form introduced in the previous chapter. Some further worth-mentioning points include first and foremost that the system structure considered here is, in general, multivariable. Moreover, constant matrix $\mathbf{F}_2$ is actually a generalization of constant matrix A_{OI} in the Brunovsky canonical form [9–19] considered in the previous chapter. Integer parameter "rho-two" (ρ_2) is an alternative to relative degree introduced in the previous chapter; it is conveniently applicable to the affine subsystem of the overall nonlinear system with many inputs and outputs we are considering. Matrix block $\mathbf{F}_1$ is a modified structural block holding the place of row matrix $\underline{a}$ in the Brunovsky canonical form [9–19] considered in the previous chapter. Finally, as can be seen, matrix $\mathbf{G}$ obtains a form similar to that of the product ($\mathrm{b}_\mathrm{I}\beta^{-1}$) arising, e.g., in input-state linearization in the previous chapter; more specifically, square block $\mathbf{G}_1(\mathbf{x}_1)$ takes the place of $\beta^{-1}(\mathbf{z})$ in the general SISO exact input-state linearization framework.

Then, we demonstrate that if the following control input is applied to the system, the BP subsystem becomes linear in its state and with respect to the synthetic input vector $\mathbf{v}$ of dimension m_2, i.e., the same as $\mathbf{u}$:

$$\mathbf{u} = \mathbf{G}_1^{-1}(\mathbf{x}_1)\cdot(\mathbf{v} - \mathbf{F}_1(\mathbf{x}_1)\cdot\mathbf{x}_2). \tag{3.17}$$

Indeed:

$$\begin{aligned}
\dot{\mathbf{x}}_2 &= \mathbf{F}(\mathbf{x}_1)\cdot\mathbf{x}_2 + \mathbf{G}(\mathbf{x}_1)\cdot\mathbf{u} \\
&= \begin{bmatrix} \mathbf{F}_2 \\ \hline\hline \mathbf{F}_1(\mathbf{x}_1) \end{bmatrix}\cdot\mathbf{x}_2 + \begin{bmatrix} \mathbf{0}_{(n_2-m_2)\times m_2} \\ \hline\hline \mathbf{G}_1(\mathbf{x}_1) \end{bmatrix}\cdot\mathbf{G}_1^{-1}(\mathbf{x}_1)\cdot(\mathbf{v}-\mathbf{F}_1(\mathbf{x}_1)\cdot\mathbf{x}_2) \\
&= \begin{bmatrix} \mathbf{F}_2\cdot\mathbf{x}_2 \\ \mathbf{F}_1(\mathbf{x}_1)\cdot\mathbf{x}_2 \end{bmatrix} + \begin{bmatrix} \vec{\mathbf{0}}_{(n_2-m_2)} \\ \mathbf{v}-\mathbf{F}_1(\mathbf{x}_1)\cdot\mathbf{x}_2 \end{bmatrix} = \begin{bmatrix} \mathbf{F}_2\cdot\mathbf{x}_2 \\ \mathbf{v} \end{bmatrix}.
\end{aligned}$$

So, in effect, the BP subsystem equations after exact feedback linearization is applied according to the procedure presented become as follows:

$$\begin{aligned}
\dot{\mathbf{x}}_2 &= \begin{bmatrix} \mathbf{F}_2 \\ \hline\hline \mathbf{0}_{m_2\times n_2} \end{bmatrix}\cdot\mathbf{x}_2 + \begin{bmatrix} \mathbf{0}_{(n_2-m_2)\times m_2} \\ \hline\hline \mathbf{I}_{m_2\times m_2} \end{bmatrix}\cdot\mathbf{v} = \mathbf{\Theta}\mathbf{x}_2 + \mathbf{B}_{0\mathrm{I}}\mathbf{v} \\
\mathbf{y}_2 &= \mathbf{C}_2\cdot\mathbf{x}_2.
\end{aligned} \tag{3.18}$$

Note that the system is automatically linearized in the input-output sense too since the output equation in the above is linear in the first place by assumption.

3.4 EXACT LINEARIZATION OF THE LOW-PASS EQUIVALENT

We now proceed and extend the exact linearization method established in the previous section for the BP subsystem to its LP equivalent. We will consider the pre-envelope and the complex envelope for the BP subsystem's state and input vector; when we first introduced these concepts in Chapter 1 we saw that by using the pre-envelope, the complex envelope can be defined for a BP signal vector with carrier frequency $\omega_c = 2\pi f_c$:

$$\begin{aligned}
\mathbf{x}_{2+}(t) &= \tilde{\mathbf{x}}_2(t)\exp(j\omega_c t) \\
\mathbf{u}_{+}(t) &= \tilde{\mathbf{u}}(t)\exp(j\omega_c t).
\end{aligned} \tag{3.19}$$

As explained in Chapter 1, the pre-envelope is a complex signal vector allowed to have nonzero spectrum only in non-negative frequencies. Furthermore, as we saw earlier in Chapter 1 a similar pattern can be applied to the output vector of the BP subsystem as follows:

$$\left.\begin{aligned}
\mathbf{y}_{2+}(t) &= \tilde{\mathbf{y}}_2(t)\exp(j\omega_c t) \\
\mathbf{y}_2(t) &= \mathrm{Re}\{\mathbf{y}_{2+}(t)\}
\end{aligned}\right\} \Rightarrow \tilde{\mathbf{y}}_2 = \mathbf{C}_2\cdot\tilde{\mathbf{x}}_2. \tag{3.20}$$

The last one of the above relationships in combination with the fact that signals $\mathbf{u}$, $\mathbf{x}_2$ and consequently $\mathbf{y}_2$ are BP with carrier frequency $\omega_c = 2\pi f_c$ leads to the conclusion that their complex are LP signals. So for the (LP) complex envelopes the BP subsystem equations become the LP equivalent system of Equations (3.21)

$$\begin{aligned}\dot{\tilde{\mathbf{x}}}_2 &= [\mathbf{F}(\mathbf{x}_1) - j\omega_c \mathbf{I}] \cdot \tilde{\mathbf{x}}_2 + \mathbf{G}(\mathbf{x}_1) \cdot \tilde{\mathbf{u}} \\ \tilde{\mathbf{y}}_2 &= \mathbf{C}_2 \cdot \tilde{\mathbf{x}}_2 .\end{aligned} \tag{3.21}$$

We then proceed to assume as in the previous section that matrices $\mathbf{F}$ and $\mathbf{G}$ have the following form:

$$\mathbf{F}(\mathbf{x}_1) = \left[\begin{array}{c} \mathbf{F}_2 \\ \hline \hline \mathbf{F}_1(\mathbf{x}_1) \end{array}\right], \quad \mathbf{G}(\mathbf{x}_1) = \left[\begin{array}{c} \mathbf{0}_{(n_2-m_2)\times m_2} \\ \hline \hline \mathbf{G}_1(\mathbf{x}_1) \end{array}\right]. \tag{3.22}$$

Again:

$\rho_2 = n_2 - m_2$, i.e., the excess in number of states vs. inputs of the BP subsystem.

$\mathbf{F}_2$ is a $\rho_2 \times n_2$ constant matrix.

$\mathbf{F}_1$ is a $m_2 \times n_2$ matrix function of LP subsystem state vector $\mathbf{x}_1$.

$\mathbf{0}$ is the zero matrix or vector of specified dimensions.

$\mathbf{G}_1$ is a $m_2 \times m_2$ matrix function of LP subsystem state vector $\mathbf{x}_1$; $\mathbf{G}_1$ must be invertible in at least a domain $\mathbf{D}_1 \subset \mathbb{R}^{\kappa}$ where $\kappa = n_1$ is the dimension of the LP subsystem state vector $\mathbf{x}_1$.

We now evaluate the form of the system when the control input $\mathbf{u}$ is chosen as follows:

$$\left.\begin{array}{c} \mathbf{u} = \mathbf{G}_1^{-1}(\mathbf{x}_1) \cdot (\mathbf{v} - \mathbf{F}_1(\mathbf{x}_1) \cdot \mathbf{x}_2) \\ \Updownarrow \\ \hat{\mathbf{u}} = \mathbf{G}_1^{-1}(\mathbf{x}_1) \cdot (\hat{\mathbf{v}} - \mathbf{F}_1(\mathbf{x}_1) \cdot \hat{\mathbf{x}}_2) \end{array}\right\} \Rightarrow \mathbf{u}_+ = \mathbf{G}_1^{-1}(\mathbf{x}_1) \cdot (\mathbf{v}_+ - \mathbf{F}_1(\mathbf{x}_1) \cdot \mathbf{x}_{2+}) . \tag{3.23}$$

Then, we demonstrate that by employing the control input above, the BP subsystem's LP equivalent becomes linear in its state's complex envelope vector and with respect to the synthetic input's complex envelope vector. As before, the synthetic input $\mathbf{v}$ has dimension m_2, i.e., the same as $\mathbf{u}$. Furthermore for its Hilbert Transform, pre-envelope and complex envelope the following hold as derived in detail in Chapter 1:

$$\hat{\mathbf{v}}(t) \triangleq \frac{1}{\pi} \int_{-\infty}^{+\infty} \frac{1}{t - t_1} \mathbf{v}(t_1) dt_1 = \frac{1}{\pi t} * \mathbf{v}(t) \Leftrightarrow \hat{\mathbf{v}}(f) = -j \ \mathrm{sgn}(f) \mathbf{v}(f)$$

$$\mathbf{v}_+(t) \triangleq \mathbf{v}(t) + j\hat{\mathbf{v}}(t) = \tilde{\mathbf{v}}(t) \exp(j\omega_c t) \Leftrightarrow \tilde{\mathbf{v}}(t) \triangleq (\mathbf{v}(t) + j\hat{\mathbf{v}}(t)) \exp(-j\omega_c t).$$

The complex envelope of synthetic input $\mathbf{v}$ may be decomposed to a real and an imaginary component. In telecommunications literature the real component, $\mathbf{v}_C(t)$, is referred to as the in-phase (or **I** for short) component and the imaginary component, $\mathbf{v}_S(t)$, is referred to as the quadrature (or Q for short) component. Clearly, the I and Q components of the complex envelope are mutually orthogonal and preserve the complete information content of the BP signal from which they are generated. Furthermore, as can be seen from the first one of Equations (1.22), the complex envelope is a generalization of the concept of amplitude modulation applied to the generalized imaginary exponential carrier signal $\exp(j\omega_c t)$:

$$\tilde{\mathbf{v}}(t) = \mathbf{v}_C(t) + j\mathbf{v}_S(t). \tag{3.24}$$

Finally, it is noted that the spectrum of the complex envelope of the synthetic input $\mathbf{v}$ is LP as expected. Indeed, for the Fourier Transform of $\mathbf{v}$ the following can be derived:

$$\tilde{\mathbf{v}}(f) = \mathbf{v}_+(f - f_c) = 2\mathbf{v}(f - f_c). \tag{3.25}$$

We proceed now to use the complex envelope of the control input in Equation (3.23):

$$\begin{gathered}\tilde{\mathbf{u}}e^{j\omega_c t} = \mathbf{G}_1^{-1}(\mathbf{x}_1)\cdot\left(\tilde{\mathbf{v}}e^{j\omega_c t} - \mathbf{F}_1(\mathbf{x}_1)\cdot\tilde{\mathbf{x}}_2 e^{j\omega_c t}\right)\\ \Updownarrow \\ \tilde{\mathbf{u}} = \mathbf{G}_1^{-1}(\mathbf{x}_1)\cdot(\tilde{\mathbf{v}} - \mathbf{F}_1(\mathbf{x}_1)\cdot\tilde{\mathbf{x}}_2).\end{gathered} \tag{3.26}$$

We now substitute the above in Equation (3.21) and proceed as follows:

$$\begin{aligned}\dot{\tilde{\mathbf{x}}}_2 &= [\mathbf{F}(\mathbf{x}_1) - j\omega_c\mathbf{I}]\cdot\tilde{\mathbf{x}}_2 + \mathbf{G}(\mathbf{x}_1)\cdot\tilde{\mathbf{u}}\\ &= \left(\begin{bmatrix}\mathbf{F}_2\\ \hline\hline \mathbf{F}_1(\mathbf{x}_1)\end{bmatrix} - j\omega_c\mathbf{I}\right)\cdot\tilde{\mathbf{x}}_2 + \begin{bmatrix}\mathbf{0}_{(n_2-m_2)\times m_2}\\ \hline\hline \mathbf{G}_1(\mathbf{x}_1)\end{bmatrix}\cdot\mathbf{G}_1^{-1}(\mathbf{x}_1)\cdot(\tilde{\mathbf{v}} - \mathbf{F}_1(\mathbf{x}_1)\cdot\tilde{\mathbf{x}}_2)\\ &= \begin{bmatrix}\mathbf{F}_2\cdot\tilde{\mathbf{x}}_2\\ \mathbf{F}_1(\mathbf{x}_1)\cdot\tilde{\mathbf{x}}_2\end{bmatrix} - j\omega_c\tilde{\mathbf{x}}_2 + \begin{bmatrix}\vec{\mathbf{0}}_{(n_2-m_2)}\\ \tilde{\mathbf{v}} - \mathbf{F}_1(\mathbf{x}_1)\cdot\tilde{\mathbf{x}}_2\end{bmatrix} = \begin{bmatrix}\mathbf{F}_2\cdot\tilde{\mathbf{x}}_2\\ \tilde{\mathbf{v}}\end{bmatrix} - j\omega_c\tilde{\mathbf{x}}_2.\end{aligned}$$

In effect the following linearized LP equivalent is obtained:

$$\begin{gathered}\dot{\tilde{\mathbf{x}}}_2 = \begin{bmatrix}\mathbf{F}_2\\ \hline\hline \mathbf{0}_{m\times n}\end{bmatrix}\cdot\tilde{\mathbf{x}}_2 - j\omega_c\tilde{\mathbf{x}}_2 + \begin{bmatrix}\mathbf{0}_{(n-m)\times m}\\ \hline\hline \mathbf{I}_{m\times m}\end{bmatrix}\cdot\tilde{\mathbf{v}} = [\mathbf{\Theta} - j\omega_c\mathbf{I}]\,\tilde{\mathbf{x}}_2 + \mathbf{B}_{0\mathrm{I}}\tilde{\mathbf{v}} \Rightarrow\\ \Downarrow\\ \begin{Bmatrix}\dot{\tilde{\mathbf{x}}}_2 = \mathbf{\Theta}_{LP}\tilde{\mathbf{x}}_2 + \mathbf{B}_{0\mathrm{I}}\tilde{\mathbf{v}}\\ \tilde{\mathbf{y}}_2 = \mathbf{C}_2\cdot\tilde{\mathbf{x}}_2\end{Bmatrix}.\end{gathered} \tag{3.27}$$

The above means that if the original BP subsystem is amenable to exact feedback linearization so is its LP equivalent at least in the manner presented in this chapter.

CHAPTER 4

Electromechanical System Applications

4.1 ELECTROMECHANICAL SYSTEM GOVERNING EQUATIONS

We return now to the system introduced in Chapter 1 that exhibits nonlinear cross-band coupling and state modulation and demodulation. By using first principles for the system of Fig. 1.2, the following Lagrangian, $\mathbf{L}_L$, is obtained for the non-dissipative and unforced case $(R = 0, b = 0$, and $e = 0)$ [20–24]:

$$\mathrm{L}_L(q, \dot{q}, y, \dot{y}) = \frac{1}{2} L(y) \dot{q}^2 + \frac{1}{2} m \dot{y}^2 - \frac{1}{2C} q^2 - \frac{1}{2} k y^2. \tag{4.1}$$

In the above, $q(t)$ denotes the capacitor's charge and, therefore, $\dot{q}(t) = i(t)$ is the circuit's current; $x(t), \dot{x}(t)$ are the payload displacement position and velocity, respectively; L is the electromagnet's inductance, R and C the circuit resistance and capacitance; m is the payload mass, b the damping coefficient, and k the spring constant of the mechanical oscillator. Then, we introduce the following canonical coordinates [20–24]:

$$p_y = \frac{\partial \mathrm{L}_L}{\partial \dot{y}} = m \dot{y} \tag{4.2}$$

$$p_q = \frac{\partial \mathrm{L}_L}{\partial \dot{q}} = L(y) \cdot \dot{q}. \tag{4.3}$$

In result, the Hamiltonian, H, is obtained as follows [20–24]:

$$\mathrm{H}(q, p_q, y, p_y) = \frac{1}{2} \cdot \frac{p_q^2}{L(y)} + \frac{1}{2} \cdot \frac{p_y^2}{m} + \frac{1}{2C} \cdot q^2 + \frac{1}{2} \cdot k y^2. \tag{4.4}$$

The Hamiltonian describes the non-dissipative system. For the actual system, including the damping terms and excitation (forcing) the following set of equations is derived [20–24]:

$$i \triangleq \dot{q} = \frac{p_q}{L(y)} \tag{4.5}$$

$$\dot{y} = \frac{p_y}{m} \tag{4.6}$$

$$\dot{p}_q = -\frac{1}{C} \cdot q - p_q \cdot \frac{R}{L(y)} + e \tag{4.7}$$

$$\dot{p}_y = -ky - p_y \frac{b}{m} + \frac{1}{2} \cdot \frac{p_q^2}{L^2(y)} \cdot \frac{dL}{dy}. \tag{4.8}$$

In effect, one obtains the following second-order dynamical equations for the electrical and mechanical subsystems:

$$L_1 [\dot{y}(t) \cdot \dot{q}(t) + y(t) \cdot \ddot{q}(t)] + L_0 \ddot{q}(t) + R\dot{q}(t) + \frac{1}{C} q(t) = e(t) \tag{4.9}$$

$$m\ddot{y}(t) + b\dot{y}(t) + ky(t) = \frac{dL}{dy} \cdot \frac{i^2(t)}{2}. \tag{4.10}$$

The above system consists of two coupled second-order oscillators. However, the coupling is nonlinear. Indeed, the right-hand magnetic force term of Equation (4.10) and the first term of the left-hand side of Equation (4.9) are clearly nonlinear.

Furthermore, if the inductance were not a function of the payload displacement, then coupling would not take place. A common dependence of the inductance on payload displacement that can be justified by electromechanical theory and analysis of magnetic circuits is the following [20–24]:

$$L(y) = L_0 + L_1 y, \quad 0 \le y \le y_{em}. \tag{4.11}$$

The equation above is valid only a limited finite range of mass displacement y. It comes only as an approximation to the commonly encountered sigmoid dependence [20–24], like, e.g., that depicted by the logistic sigmoid, of inductance to the position of a metallic mass like the one considered here. In specific, a relationship for $L(y)$ valid over the entire real value range for y might look like the following:

$$L(y) = L_\infty + \frac{(\chi - 1) L_\infty}{1 + \exp\left[\frac{2 y_{em} L_1}{(\chi - 1) L_\infty} \left(1 - 2\frac{y}{y_{em}}\right)\right]}, \quad \chi > 1. \tag{4.12}$$

Based on the above one obtains the following facts:

$$\begin{aligned}
&L(y \to -\infty) = L_\infty, \quad L(y \to +\infty) = \chi L_\infty, \\
&L\left(\frac{y_{em}}{2}\right) = \frac{\chi + 1}{2} L_\infty \\
&L(0) = L_\infty + \frac{(\chi - 1) L_\infty}{1 + \exp\left(\frac{2 y_{em} L_1}{(\chi - 1) L_\infty}\right)} \\
&L(y_{em}) = L_\infty + \frac{(\chi - 1) L_\infty}{1 + \exp\left(-\frac{2 y_{em} L_1}{(\chi - 1) L_\infty}\right)}.
\end{aligned} \tag{4.13}$$

By applying Taylor's expansion to Equation (4.12) around point and keeping only the first-order term one obtains the following:

$$L(y) = \frac{\chi+1}{2}L_\infty + L_1\left(y - \frac{y_{em}}{2}\right) = \underbrace{\frac{(\chi+1)\,L_\infty - L_1 y_{em}}{2}}_{L_0} + L_1 y. \tag{4.14}$$

Without any loss of generality assume that the electromagnet in Fig. 1.2 is placed at a position on the y-axis equal to the characteristic length y_{em}. The characteristic length can be calculated as the distance between the position where the mass is inductively decoupled from the electromagnet and the position where it has the maximum effect on the inductance.

$$y_{em} = \frac{L(+\infty) - L(-\infty)}{L_1} = \frac{\chi-1}{L_1}L_\infty. \tag{4.15}$$

Using the above, the following observations should be noted:

$$L(y_{em}) \approx L(+\infty), \quad L(0) \approx L(-\infty), \quad L_0 = L_\infty$$
$$L(y) \simeq \begin{cases} L_\infty, & y < 0 \\ L_\infty + L_1 y, & 0 \le y \le y_{em} \\ L_\infty + L_1 y_{em}, & y > y_{em}. \end{cases} \tag{4.16}$$

So, provided that the mass displacement is constrained within the interval indicated in Equation (4.11), the central branch of the equation above can be used. In the system that will be considered in the remaining of this text, hard stoppers will be employed to ensure that the mass displacement remains within the permissible range. Furthermore, for the sake of simplicity and without loss of generality, the electromagnet will be placed at position y_{em} and the natural length of the mechanism's spring will correspond to $y = 0$, while y is assumed to increase from the spring's natural length toward the electromagnet. By using Equation (4.11) in Equations (4.9) and (4.10) one obtains:

$$L_1\left[\dot{y}(t)\cdot\dot{q}(t) + y(t)\cdot\ddot{q}(t)\right] + L_0\ddot{q}(t) + R\dot{q}(t) + \frac{1}{C}q(t) = e(t) \tag{4.17}$$

$$m\ddot{y}(t) + b\dot{y}(t) + ky(t) = \frac{L_1}{2}i^2(t), \quad 0 \le y \le y_{em}. \tag{4.18}$$

The nonlinear coupling between the two oscillators is quantified clearly in (4.12) and (4.13). Furthermore, as can be seen decoupling occurs if $L_1 = 0$.

4.2 FORMULATION OF SYSTEM DYNAMICS IN STATE SPACE

As seen in the previous section, for the analysis of the system of coupled electromechanical oscillators shown in Fig. 1.2, a few basic principles were employed to obtain the following set of nonlinear, second-order, ordinary differential equations for payload displacement y and capacitor charge q:

$$m\ddot{y} + b\dot{y} + ky = \frac{L_1}{2}i^2 + d \tag{4.19}$$

$$[L_0 + L_1 y]\ddot{q} + [R + L_1\dot{y}]\dot{q} + \frac{1}{C}q = e. \tag{4.20}$$

In the above, the force disturbance signal d has been superimposed to the electromagnet's force in the LP mechanical subsystem. As it is common practice for electromechanical systems the following state vector may be used:

$$\mathbf{x} = \begin{bmatrix} y & \dot{y} & q & \dot{q} \end{bmatrix}^T. \tag{4.21}$$

This is partitioned as follows for establishing the correspondence with the LP-BP formulation presented earlier in Chapter 1 and in the previous section:

$$\mathbf{x}_1 = \begin{bmatrix} y \\ \dot{y} \end{bmatrix}, \quad \mathbf{d} = \begin{bmatrix} 0 \\ \dfrac{d}{m} \end{bmatrix}, \quad \mathbf{x}_2 = \begin{bmatrix} q \\ \dot{q} \end{bmatrix}, \quad \mathbf{u} = e, \quad \mathbf{y}_2 = i. \tag{4.22}$$

By using the above, the following matrices are obtained for the BP-LP state space decomposition when applied to the case of the coupled electromechanical oscillators at hand:

$$\begin{gathered}
\mathbf{A}_{LP} = \begin{bmatrix} 0 & 1 \\ -\dfrac{k}{m} & -\dfrac{b}{m} \end{bmatrix}, \quad \boldsymbol{\psi}(\mathbf{y}_2) = \begin{bmatrix} 0 \\ \dfrac{L_1}{2m}i^2 \end{bmatrix}, \\
\mathbf{F}(\mathbf{x}_1) = \begin{bmatrix} 0 & 1 \\ -\dfrac{1}{(L_0 + L_1 y)C} & -\dfrac{R + L_1\dot{y}}{L_0 + L_1 y} \end{bmatrix}, \\
\mathbf{G}(\mathbf{x}_1) = \begin{bmatrix} 0 \\ \dfrac{1}{L_0 + L_1 y} \end{bmatrix}, \quad \mathbf{C}_2 = \begin{bmatrix} 0 & 1 \end{bmatrix}.
\end{gathered} \tag{4.23}$$

As can be seen in this case, $\boldsymbol{\psi}(\mathbf{y}_2)$ is already in the required expansion form specified in Chapters 1 and 2; therefore, no further treatment is needed. Otherwise, transfer function matrix $\mathbf{H}_1(s) = (s\mathbf{I} - \mathbf{A})^{-1}$ is given by the following: transfer function matrix $\mathbf{H}_1(s) = (s\mathbf{I} - \mathbf{A})^{-1}$ is given by the following:

$$\mathbf{H}_1(s) = \frac{1}{ms^2 + bs + k} \begin{bmatrix} (ms+b) & m \\ k & ms \end{bmatrix}. \tag{4.24}$$

The poles of the above are the roots of (characteristic) polynomial $P_1(s) = ms2 + bs + k$. It can also be seen that:

$$\mathbf{H}_1(s = j0) = \begin{bmatrix} \frac{b}{k} & \frac{m}{k} \\ 1 & 0 \end{bmatrix}. \tag{4.25}$$

For frequency $s = j\omega$ going to infinity, one can straightforwardly verify that all scalar transfer functions in the entries of transfer function matrix $\mathbf{H}_1(s)$ vanish. Therefore, the LP requirement for $\mathbf{H}_1(s)$ clearly checks out. In a more general case, however, one should employ Singular Value Decomposition as outlined in Chapter 1 in order to establish whether $\mathbf{H}_1(s)$ is LP or not.

We now need to proceed to establish whether transfer function matrix $\mathbf{H}_2(s) = \mathbf{C}_2(s\mathbf{I} - \mathbf{F}_0)^{-1}\boldsymbol{\Gamma}$ is BP or not as explained in Chapter 1. In this respect, we need to calculate matrices $\mathbf{F}_0$ and $\boldsymbol{\Gamma}$. This can be achieved by employing their Taylor expansion as outlined in Chapter 1. For the matrix Taylor expansions, though, the following scalar one is useful:

$$\frac{1}{L_0 + L_1 y} = \frac{1}{L_0} - \frac{L_1}{L_0^2} y + \frac{L_1^2}{L_0^3} y^2 - \frac{L_1^3}{L_0^4} y^3 + \cdots \tag{4.26}$$

Furthermore, the above yields:

$$\frac{1}{L_0 + L_1 y} = \frac{1}{L_0\left(1 + \frac{L_1}{L_0} y\right)} = \frac{1}{L_0}\left(1 - \frac{L_1}{L_0} y + \frac{L_1^2}{L_0^2} y^2 - \frac{L_1^3}{L_0^3} y^3 + \cdots\right). \tag{4.27}$$

For micro- (or even nano-) electromechanical systems (MEMS or NEMS) [6, 10] as well as in other applied mechatronics [4, 5] it is reasonable to assume that $L_1 y << L_0$. With this in mind, the following first-order approximation will be considered in the above:

$$\frac{1}{L_0 + L_1 y} \cong \frac{1}{L_0}\left(1 - \frac{L_1}{L_0} y\right) = \frac{1}{L_0} - \frac{L_1}{L_0^2} y. \tag{4.28}$$

Finally, by employing the above one obtains the following for multivariable matrix functions $\mathbf{F}$ and $\mathbf{G}$:

$$\mathbf{F}(\mathbf{x}_1) = \underbrace{\begin{bmatrix} 0 & 1 \\ -\frac{1}{CL_0} & -\frac{R}{L_0} \end{bmatrix}}_{\mathbf{F}_0} + \underbrace{\begin{bmatrix} 0 & 0 \\ \frac{L_1}{CL_0^2} y & \begin{bmatrix} \frac{RL_1}{L_0^2} & -\frac{L_1}{L_0} \end{bmatrix} \cdot \begin{bmatrix} y \\ \dot{y} \end{bmatrix} \end{bmatrix}}_{\mathbf{F}_1\left(\mathbf{x}_1^{\otimes 1}\right)} \tag{4.29}$$

$$\mathbf{G}(\mathbf{x}_1) = \underbrace{\begin{bmatrix} 0 \\ \frac{1}{L_0} \end{bmatrix}}_{\mathbf{G}_0} + \underbrace{\begin{bmatrix} 0 \\ -\frac{L_1}{L_0^2} y \end{bmatrix}}_{\mathbf{G}_1\left(\mathbf{x}_1^{\otimes 1}\right)}. \tag{4.30}$$

By using the above one obtains the following for matrix $\boldsymbol{\Gamma}$:

$$\boldsymbol{\Gamma} = \begin{bmatrix} \mathbf{G}_0 & \boldsymbol{\Gamma}_0 \end{bmatrix} = \begin{bmatrix} 0 & 0 & 0 \\ \frac{1}{L_0} & 1 & 1 \end{bmatrix}. \tag{4.31}$$

Notice that the above form for $\boldsymbol{\Gamma}$ as well the one for $\boldsymbol{\Gamma}_0$ does not change even if the approximation in Equation (4.28) is dropped. So in result the following is obtained for transfer function matrix $\mathbf{H}_2(s)$:

$$\mathbf{H}_2(s) = \mathbf{C}_2 \cdot (s\mathbf{I} - \mathbf{F}_0)^{-1} \cdot \boldsymbol{\Gamma} = \frac{Cs}{L_0 C s^2 + RCs + 1} \begin{bmatrix} 1 & L_0 & L_0 \end{bmatrix}. \tag{4.32}$$

The poles of the above are the roots of (characteristic) polynomial $P_2(s) = L_0 C s^2 + RCs + 1$. It can also be seen that:

$$\mathbf{H}_2(s = j0) = \begin{bmatrix} 0 & 0 & 0 \end{bmatrix}. \tag{4.33}$$

Also, as frequency $s = j\omega$ is going to infinity, one can straightforwardly verify that all scalar transfer functions in the entries of transfer function matrix $\mathbf{H}_2(s)$ vanish. Actually, only in a vicinity of frequency ω_E, defined below, the elements of $\mathbf{H}_2(s)$ assume magnitude-wise non-negligible values. Therefore, the BP requirement with carrier frequency $\omega_c = \omega_E$ for $\mathbf{H}_2(s)$ is therefore established:

$$\omega_c = \omega_E = \sqrt{\frac{1}{CL_0}}. \tag{4.34}$$

The values of the entries in $\mathbf{H}_2(s)$ when $\omega = \omega_c = \omega_E$ are given by the following:

$$\mathbf{H}_2(j\omega_E) = \frac{1}{R} \begin{bmatrix} 1 & L_0 & L_0 \end{bmatrix}. \tag{4.35}$$

In summary, the LP equivalent of the system consisting of the two coupled electromechanical oscillators at hand is given by the following:

$$\begin{bmatrix} \dot{y} \\ \ddot{y} \end{bmatrix} = \begin{bmatrix} 0 & 1 \\ -\frac{k}{m} & -\frac{b}{m} \end{bmatrix} \cdot \begin{bmatrix} y \\ \dot{y} \end{bmatrix} + \begin{bmatrix} 0 \\ \frac{L_1}{2m} \cdot \frac{|\tilde{i}|^2}{2} \end{bmatrix} + \begin{bmatrix} 0 \\ \frac{d}{m} \end{bmatrix} \tag{4.36}$$

$$\frac{d}{dt} \begin{bmatrix} \tilde{q} \\ \tilde{i} \end{bmatrix} = \begin{bmatrix} -j\omega_E & 1 \\ -\frac{1}{(L_0 + L_1 y) C} & -\frac{R + L_1 \dot{y}}{L_0 + L_1 y} - j\omega_E \end{bmatrix} \cdot \begin{bmatrix} \tilde{q} \\ \tilde{i} \end{bmatrix} + \begin{bmatrix} 0 \\ \frac{1}{L_0 + L_1 y} \end{bmatrix} \tilde{e} \tag{4.37}$$

$$\tilde{i} = \dot{\tilde{q}} = \frac{d\tilde{q}}{dt} + j\omega_E\tilde{q} \Rightarrow \frac{d\tilde{q}}{dt} = -j\omega_E\tilde{q} + \tilde{i}. \tag{4.38}$$

The above is the full LP equivalent system, meaning that matrices **F** and **G** are used in full and not by some approximation. If the first-order approximations in Equations (4.29) and (4.30) are employed instead in the dynamics of the LP equivalent in Equation (4.37) perturbation analysis may be carried out [1, 20, 21]. Perturbation analysis clearly reveals that by using the LP equivalent system the solutions obtained are identical for the electromechanical coupled oscillators at hand [1, 20, 21]. However, in the remainder of this chapter we intend to apply the modified exact feedback linearization approach introduced in Chapter 3 for nonlinear systems with state modulation. We will then run numerical simulations to confirm the approach.

4.3 EXACT FEEDBACK LINEARIZATION OF THE ELECTROMECHANICAL SYSTEM

We now proceed and apply the feedback linearization approach to the BP subsystem of the coupled electromechanical oscillators, which is no other than the electrical part of the system. We recall here that:

$$\mathbf{A}_{LP} = \begin{bmatrix} 0 & 1 \\ -\frac{k}{m} & -\frac{b}{m} \end{bmatrix}, \quad \boldsymbol{\psi}(\mathbf{y}_2) = \begin{bmatrix} 0 \\ \frac{L_1}{2m} i^2 \end{bmatrix},$$

$$\mathbf{F}(\mathbf{x}_1) = \begin{bmatrix} 0 & 1 \\ -\frac{1}{(L_0 + L_1 y)C} & -\frac{R + L_1\dot{y}}{L_0 + L_1 y} \end{bmatrix}, \tag{4.39}$$

$$\mathbf{G}(\mathbf{x}_1) = \begin{bmatrix} 0 \\ \frac{1}{L_0 + L_1 y} \end{bmatrix}, \quad \mathbf{C}_2 = \begin{bmatrix} 0 & 1 \end{bmatrix}.$$

Then, the following can be derived concerning the required structure toward BP linearization:

$$n_2 = 2, \quad m_2 = 1, \quad \rho_2 = n_2 - m_2 = 1$$

$$\mathbf{F}_2 = \begin{bmatrix} 0 & 1 \end{bmatrix}, \quad \mathbf{F}_1(\mathbf{x}_1) = \begin{bmatrix} -\frac{1}{(L_0 + L_1 y)C} & -\frac{R + L_1\dot{y}}{L_0 + L_1 y} \end{bmatrix}, \tag{4.40}$$

$$\mathbf{G}_1(\mathbf{x}_1) = \frac{1}{L_0 + L_1 y}.$$

In effect, one obtains that:

$$\mathbf{F}(\mathbf{x}_1) = \left[\begin{array}{cc} 0 & 1 \\ \hline\hline -\frac{1}{(L_0 + L_1 y)C} & -\frac{R + L_1\dot{y}}{L_0 + L_1 y} \end{array}\right], \quad \mathbf{G}(\mathbf{x}_1) = \left[\begin{array}{c} 0 \\ \hline\hline \frac{1}{L_0 + L_1 y} \end{array}\right]. \tag{4.41}$$

In effect the following control input allows the exact linearization with respect to both the states and the output of the BP subsystem:

$$\begin{gathered}\mathbf{u} = \mathbf{G}_1^{-1}(\mathbf{x}_1) \cdot (\mathbf{v} - \mathbf{F}_1(\mathbf{x}_1) \cdot \mathbf{x}_2) \\ \Downarrow \\ e = (L_0 + L_1 y)\left(\mathrm{q}_{tt} + \frac{q}{(L_0 + L_1 y)C} + \frac{(R + L_1\dot{y})\,i}{L_0 + L_1 y}\right) \\ \Downarrow \\ e = (L_0 + L_1 y)\,\mathrm{q}_{tt} + C^{-1}q + (R + L_1\dot{y})\,i.\end{gathered} \tag{4.42}$$

In the above, q_{tt} is the synthetic input with respect to which the BP subsystem is expected to be linear. Indeed, by using Equation (4.42) the BP subsystem dynamic equation become as follows:

$$\begin{gathered}L_1(\dot{y}i + y\ddot{q}) + L_0\ddot{q} + Ri + \frac{1}{C}q = (L_0 + L_1 y)\,\mathrm{q}_{tt} + C^{-1}q + (R + L_1\dot{y})\,i \\ \Downarrow \\ (L_0 + L_1 y)\,\ddot{q} = (L_0 + L_1 y)\,\mathrm{q}_{tt} \\ \Downarrow \\ \ddot{q} = \mathrm{q}_{tt}.\end{gathered} \tag{4.43}$$

In the above, synthetic input q_{tt} has units of *electric current rate*, i.e., A/s (or equivalently C/s^2) in the International System of Units (SI). The linear state equations of the BP system when the above linearizing control voltage, e, is used become just a couple of integrators connected in tandem, as follows.

$$\begin{gathered}\begin{bmatrix}\dot{q} \\ \ddot{q}\end{bmatrix} = \begin{bmatrix}0 & 1 \\ 0 & 0\end{bmatrix}\begin{bmatrix}q \\ \dot{q}\end{bmatrix} + \begin{bmatrix}0 \\ 1\end{bmatrix}\mathrm{q}_{tt} \\ i = \dot{q} \Rightarrow \frac{di}{dt} = \ddot{q} = \mathrm{q}_{tt}.\end{gathered} \tag{4.44}$$

So now the current through the circuit that controls the voltage applied on the payload of the mechanical LP subsystem is directly controlled by the synthetic input q_{tt} in a straightforward linear manner. Furthermore, notice that the linearized system is in the Brunovsky canonical form [9–19] with two integrators.

For the LP equivalent, a similar pattern arises with of course the complex envelope of the synthetic input, i.e., the complex envelope of the electric current rate, replacing q_{tt} in the equations. Specifically, for the LP equivalent of the BP subsystem the control input to be used is as follows:

$$\begin{gathered}\tilde{\mathbf{u}} = \mathbf{G}_1^{-1}(\mathbf{x}_1) \cdot (\tilde{\mathbf{v}} - \mathbf{F}_1(\mathbf{x}_1) \cdot \tilde{\mathbf{x}}_2) \\ \Downarrow \\ \tilde{e} = (L_0 + L_1 y)\left(\tilde{\mathrm{q}}_{tt} + \frac{\tilde{q}}{(L_0 + L_1 y)C} + \frac{(R + L_1\dot{y})\,\tilde{i}}{L_0 + L_1 y}\right) \\ \Downarrow \\ \tilde{e} = (L_0 + L_1 y)\,\tilde{\mathrm{q}}_{tt} + C^{-1}\tilde{q} + (R + L_1\dot{y})\,\tilde{i}.\end{gathered} \tag{4.45}$$

So now the state equations of the BP subsystem become as follows:

$$\frac{d}{dt}\begin{bmatrix}\tilde{q}\\ \tilde{i}\end{bmatrix}=\begin{bmatrix}-j\omega_E & 1\\ -\dfrac{1}{(L_0+L_1y)\,C} & -\dfrac{R+L_1\dot{y}}{L_0+L_1y}-j\omega_E\end{bmatrix}\begin{bmatrix}\tilde{q}\\ \tilde{i}\end{bmatrix}+\begin{bmatrix}0\\ \dfrac{1}{L_0+L_1y}\end{bmatrix}\left((L_0+L_1y)\,\tilde{\mathrm{q}}_{tt}+C^{-1}\tilde{q}+(R+L_1\dot{y})\,\tilde{i}\right). \tag{4.46}$$

In effect one obtains the following:

$$\begin{bmatrix}\dfrac{d\tilde{q}}{dt}\\ \dfrac{d\tilde{i}}{dt}\end{bmatrix}=\begin{bmatrix}-j\omega_E\tilde{q}+\tilde{i}\\ -\dfrac{\tilde{q}}{(L_0+L_1y)\,C}-\left(\dfrac{R+L_1\dot{y}}{L_0+L_1y}+j\omega_E\right)\tilde{i}\end{bmatrix}+\begin{bmatrix}0\\ \tilde{\mathrm{q}}_{tt}+\dfrac{\tilde{q}}{(L_0+L_1y)\,C}+\dfrac{(R+L_1\dot{y})\,\tilde{i}}{L_0+L_1y}\end{bmatrix}. \tag{4.47}$$

The above can be further decomposed to the following scalar linear, first-order ordinary differential equations:

$$\left.\begin{aligned}\frac{d\tilde{q}}{dt}+j\omega_E\tilde{q}&=\tilde{i}\\ \frac{d\tilde{i}}{dt}+j\omega_E\tilde{i}&=\tilde{\mathrm{q}}_{tt}\end{aligned}\right\}\Rightarrow\frac{d}{dt}\begin{bmatrix}\tilde{q}\\ \tilde{i}\end{bmatrix}=\begin{bmatrix}-j\omega_E & 1\\ 0 & -j\omega_E\end{bmatrix}\begin{bmatrix}\tilde{q}\\ \tilde{i}\end{bmatrix}+\begin{bmatrix}0\\ 1\end{bmatrix}\tilde{\mathrm{q}}_{tt}. \tag{4.48}$$

The equations above for the linearized LP equivalent are not in the Brunovsky canonical form, like for the original system when exact feedback linearization is employed, but rather in the Jordan canonical form. Both the Brunovsky form for the BP subsystem and the Jordan form for its LP equivalent reflect the original structure of the electrical subsystem as it arises in the coupled oscillators.

Equations (4.48) can also be derived independently and directly from Equations (4.44) that depict the dynamics of the original coupled system after the modified exact feedback linearization procedure has been carried out. Indeed:

$$\left.\begin{aligned}\dot{q} &= \frac{d}{dt}\mathrm{Re}\left(\tilde{q}e^{j\omega_E t}\right) = \frac{d}{dt}\mathrm{Re}\left(\tilde{q}^* e^{-j\omega_E t}\right) = \frac{d}{dt}\left(\frac{\tilde{q}e^{j\omega_E t} + \tilde{q}^* e^{-j\omega_E t}}{2}\right) \\ &= \left(\frac{d\tilde{q}}{dt} + j\omega_E \tilde{q}\right)\frac{e^{j\omega_E t}}{2} + \left(\frac{d\tilde{q}^*}{dt} - j\omega_E \tilde{q}^*\right)\frac{e^{-j\omega_E t}}{2} \\ \dot{q} &= \frac{\tilde{\dot{q}}e^{j\omega_E t} + \tilde{\dot{q}}^* e^{-j\omega_E t}}{2}, \quad \tilde{i} = \tilde{\dot{q}} \Rightarrow \dot{q} = \frac{\tilde{i}e^{j\omega_E t} + \tilde{i}^* e^{-j\omega_E t}}{2}\end{aligned}\right\}$$

$$\Rightarrow \left\{\begin{aligned}&\frac{d\tilde{q}}{dt} + j\omega_E\tilde{q} = \tilde{\dot{q}} = \tilde{i} \\ &\frac{d\tilde{q}^*}{dt} - j\omega_E\tilde{q}^* = \tilde{\dot{q}}^* = \tilde{i}^*\end{aligned}\right\} \Rightarrow \frac{d\tilde{q}}{dt} + j\omega_E\tilde{q} = \tilde{i}$$

$$\left.\begin{aligned}\frac{di}{dt} &= \ddot{q} = \mathrm{q}_{tt} = \mathrm{Re}\left(\tilde{\mathrm{q}}_{tt}e^{j\omega_E t}\right) = \frac{\tilde{\mathrm{q}}_{tt}e^{j\omega_E t} + \tilde{\mathrm{q}}_{tt}^* e^{-j\omega_E t}}{2} \\ \frac{di}{dt} &= \frac{d}{dt}\mathrm{Re}\left(\tilde{i}e^{j\omega_E t}\right) = \frac{d}{dt}\left(\frac{\tilde{i}e^{j\omega_E t} + \tilde{i}^* e^{-j\omega_E t}}{2}\right) \\ &= \left(\frac{d\tilde{i}}{dt} + j\omega_E\tilde{i}\right)\frac{e^{j\omega_E t}}{2} + \left(\frac{d\tilde{i}^*}{dt} - j\omega_E\tilde{i}^*\right)\frac{e^{-j\omega_E t}}{2}\end{aligned}\right\} \Rightarrow \left\{\begin{aligned}&\frac{d\tilde{i}^*}{dt} - j\omega_E\tilde{i}^* = \tilde{\mathrm{q}}_{tt}^* \\ &\frac{d\tilde{i}}{dt} + j\omega_E\tilde{i} = \tilde{\mathrm{q}}_{tt}.\end{aligned}\right.$$

Now by using, e.g., a Bode plot [14, 15], one can derive the following relationship between the square and the complex envelope of the electric current in the system's circuit:

$$i(t) = \frac{\tilde{i}e^{j\omega_E t} + \tilde{i}^* e^{-j\omega_E t}}{2} \Rightarrow i^2 = \frac{\overbrace{\tilde{i}^2 e^{j2\omega_E t} + \left(\tilde{i}^2\right)^* e^{-j2\omega_E t}}^{\text{High pass terms centered about } 2\omega_E} + 2\left(\tilde{i}e^{j\omega_E t}\right)\left(\tilde{i}^* e^{-j\omega_E t}\right)}{4}$$

$$\Rightarrow i^2 \equiv \frac{1}{2}\left|\tilde{i}\right|^2, \text{ in the baseband where the mechanical subsystem operates.}$$

This essentially expresses the physical fact that the magnetic force applied to the payload depends only on the power [25, 26] injected to the system by the source driving the circuit and not, e.g., the direction of the current. This fact, could lead to complications (e.g., undefined relative degree as seen in Chapter 3) if one attempts applying exact feedback linearization directly to the entire system.

Table 4.1: Test Case I settings

m (kg)	0.25
b (kg/s)	0.25
k (kg/s^2)	25
ω_M (rad/s)	10
L_0 (H)	0.05
L_1 (H/m)	0.5
y_{em} (cm)	10
C (F)	$2 \cdot 10^{-5}$
R (ohms)	1
ω_E (rad/s)	10^3

Finally, by employing the observation above in the baseband (i.e., LP [25, 26]) mechanical subsystem dynamics, one obtains the following:

$$
\begin{gathered}
m\ddot{y} + b\dot{y} + ky = \frac{L_1}{2} i^2 + d \\
\Updownarrow \\
m\ddot{y} + b\dot{y} + ky = \frac{L_1}{4} \left|\tilde{i}\right|^2 + d.
\end{gathered}
\tag{4.49}
$$

The above shows the importance and physical interpretation of both the LP equivalent and its value in the analysis of spectrally coupled subsystems through state modulation and demodulation. Also, as can be seen, exact feedback linearization in the modified fashion used here and introduced in Chapter 3 can be a very useful tool in practical applications of, e.g., driving electromechanical or mechatronic systems. Presentation of two numerical test cases and relevant simulations follow to further support and demonstrate this point.

4.4 TEST CASE I

The above methodology for the derivation of an input-output model for the modulating envelope dynamics of the electromechanical system under consideration is investigated by means of a typical example. Also, using the same test case the validity of the LP equivalent for future investigations is tested. The numerical settings for the example are given in Table 4.1.

Both a LP and a BP model have been set up. The models are shown in Fig. 4.1 as implemented in Matlab/Simulink©. The mechanical and electrical subsystems are set up separately in their state-space forms following the partition of the general system dynamics presented in detail before. Variable step integration is automatically employed by Matlab/Simulink© during

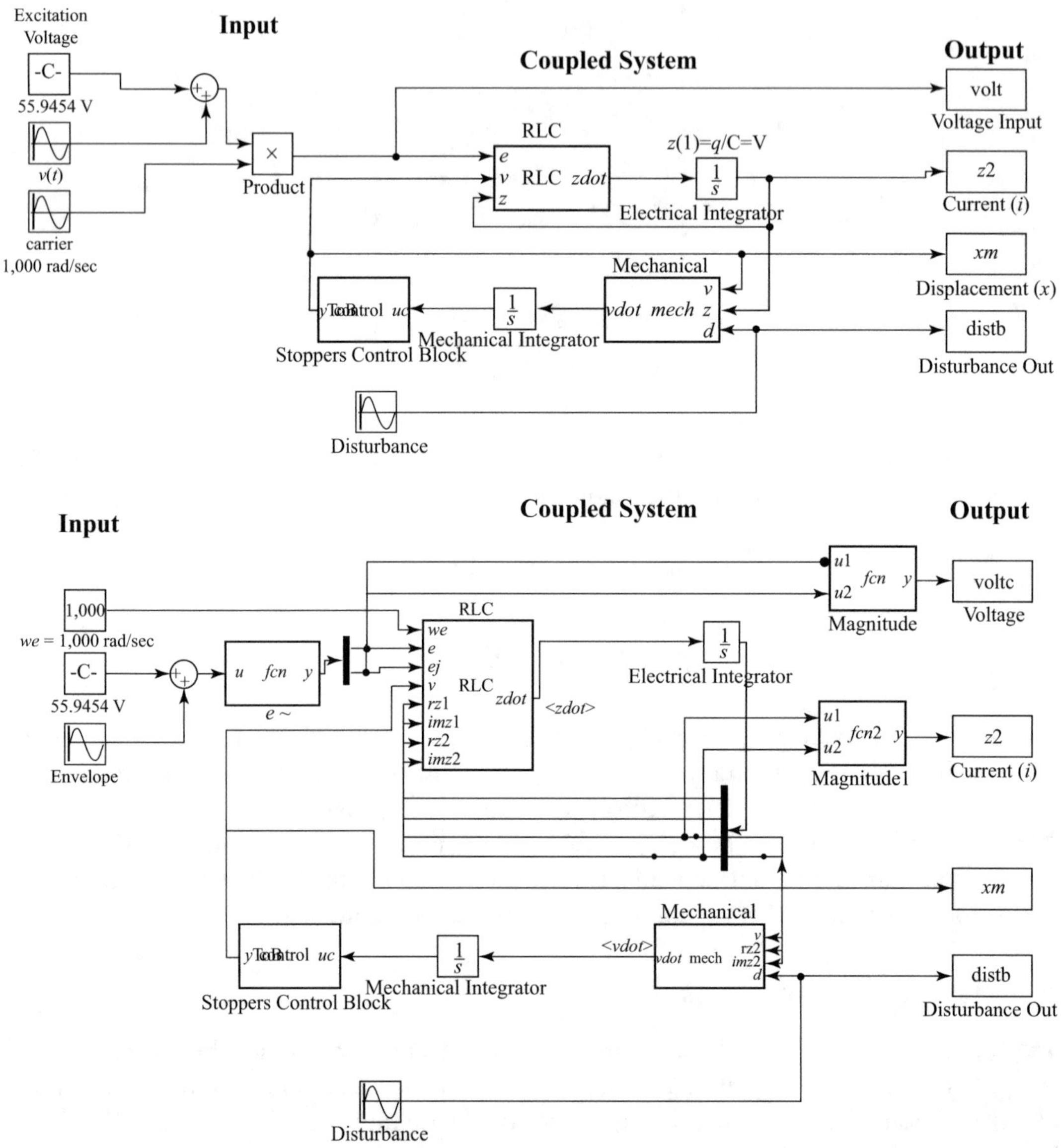

Figure 4.1: The band-pass (top) and low-pass-equivalent (bottom) model in Matlab/Simulink©.

each simulation run. Stoppers have been programmed in order to maintain the metallic mass position within the interval where the linear approximation in the equation for L holds for the electromagnet's inductance. However, by appropriately adjusting the energy input to the system the stoppers were never actually invoked.

The simplified relationship for the electromagnet's inductance is used for the simulations, instead of the accurate sigmoid one introduced in Section 4.1. This is acceptable assuming that the mass displacement remains constrained within the admissible $[0, y_{em}]$ interval. Hence the need, in both models, to introduce stoppers as a block to ensure the mass displacement remains constrained within the admissible interval.

Linearization of the equation for L is effective at point $y_{em}/2$. To implement this consistently the initial condition on displacement is set to $y_{em}/2 = 5$ cm in every simulation run with both the BP and LP model. In effect, the initial spring force will be set to $y_{em}/2 = 1.25$ N. This force needs to be counterbalanced by an electromagnet reaction force so that the mass can oscillate around $y_{em}/2$ which acts as a point of equilibrium.

This can be achieved if the voltage source's envelope demonstrates a bias offset. The value of the bias is given by the following calculation on the LP equivalent. As can be seen, the major assumption is that of undisturbed equilibrium (therefore disturbance, velocity and acceleration set to zero) around point $y_{em}/2$. Since the spring is not at its natural length, which corresponds to $y = 0$, a non-zero value of electric current envelope is needed to maintain equilibrium. Based on that required current value, the source voltage envelope value to support the equilibrium is finally calculated assuming the capacitor charge and the electric current envelopes do not change with time:

$$\begin{aligned} 0 &= \dot{y} \\ 0 &= \ddot{y} = -\frac{k}{m} \cdot \frac{y_{em}}{2} - \frac{b}{m} \cdot 0 + \frac{L_1}{2m} \cdot \frac{\left|\tilde{i}\right|^2}{2}. \end{aligned} \tag{4.50}$$

$$\begin{aligned} 0 &= \frac{d\tilde{q}}{dt} = -j\omega_E \tilde{q} + \tilde{i} \Rightarrow \tilde{i} = j\omega_E \tilde{q} \Rightarrow \tilde{q} = \frac{\tilde{i}}{j\omega_E} \\ 0 &= \ddot{\tilde{q}} = -\frac{\tilde{i}}{j\omega_E\left(L_0 + L_1\frac{y_{em}}{2}\right)C} - \left[\frac{R + L_1 \cdot 0}{L_0 + L_1\frac{y_{em}}{2}} + j\omega_E\right]\tilde{i} + \frac{\tilde{e}}{\left(L_0 + L_1\frac{y_{em}}{2}\right)}. \end{aligned} \tag{4.51}$$

Using the second unperturbed scalar equation in Equation (4.50) and setting $y_{em}/2$ as the equilibrium point one obtains the following result:

$$\left|\tilde{i}\right| = 2\sqrt{\frac{k}{L_1}y_{em}} = 2\sqrt{\frac{25}{0.5} \cdot 0.1} = 2\sqrt{5}\text{A}. \tag{4.52}$$

Then, by employing the second equation in (4.51) the value for the source voltage envelope is obtained:

$$\begin{aligned}\tilde{e} &= \left[R + j\omega_E\left(L_0 + L_1\frac{y_{em}}{2}\right) + \frac{1}{j\omega_E C}\right]\tilde{i}\\ \Rightarrow |\tilde{e}| &= \sqrt{R^2 + \left[\omega_E\left(L_0 + L_1\frac{y_{em}}{2}\right) - \frac{1}{\omega_E C}\right]^2}\,|\tilde{i}|\\ \Rightarrow |\tilde{e}| &= 111.9\,\mathrm{V} \sim 112\,\mathrm{V}.\end{aligned} \tag{4.53}$$

The same result would have been obtained if the BP model was considered instead. One needs to point out here, though, that the value obtained by Equation (4.53) corresponds to the RMS value of the BP system voltage. This is easily understood by the physics of the process, as explained in detail in Section 4.1, where the halved square of the electric current appears in the magnetic force developed on the mass.

Operation of the system with settings as in Table 4.1 is investigated by use of both the BP and LP models that were presented previously. The electrical excitation of the system (voltage envelope) was maintained constant and equal to the value calculated in Equation (4.53). On the other hand, mechanical excitation (disturbance) was applied to drive the system. In Fig. 4.2 a zero-mean, unit-variance white Gaussian noise is used after being filtered by a 16th-order Butterworth with cutoff frequency at 10 rad/s resulting to a random disturbance force in the bandwidth of the mechanical subsystem as it very well may arise in practice [25–28]. The agreement between the BP model and the LP equivalent is clearly demonstrated in the plot for metallic mass displacement, as seen in Fig. 4.2.

In Figs. 4.3 and 4.4, some of the results obtained by use of the LP and BP model are presented for three cases of disturbance force (in newton), as follows:

$$\begin{aligned}d_{\{1\}}(t) &= 1.007\cos 5t\\ d_{\{2\}}(t) &= .4098\cos 8t - .3000\sin 10t\\ d_{\{3\}}(t) &= .5035\cos 5t - .3013\sin 9t - .4075\cos 11t.\end{aligned} \tag{4.54}$$

The simulations concern the system's response to monochromatic in Fig. 4.3 and polychromatic (two and three tones) in Fig. 4.4 disturbance. Disturbance amplitudes were chosen so that the constraints for mass displacement presented earlier are not violated. The tonal frequencies on the other hand were chosen in order to excite the mechanical subsystem within its pass-band and specifically within a tight vicinity of the resonance frequency $\omega_M = 10$ rad/sec; in this manner the effect was maximized.

As seen in Figs. 4.3 and 4.4, the results are organized in three groups corresponding to the single-, dual- and triple-tone disturbance given in Equation (4.54). Each group starts with the disturbance time series applied. Then, blue corresponds to the waveforms obtained by use of the BP model while red to the LP model ones. Coincidence of the results from the two models is evident. The important feature observed there is that the imaginary part of the current LP equivalent is proportional to the mechanical oscillator mass displacement. This means that with the

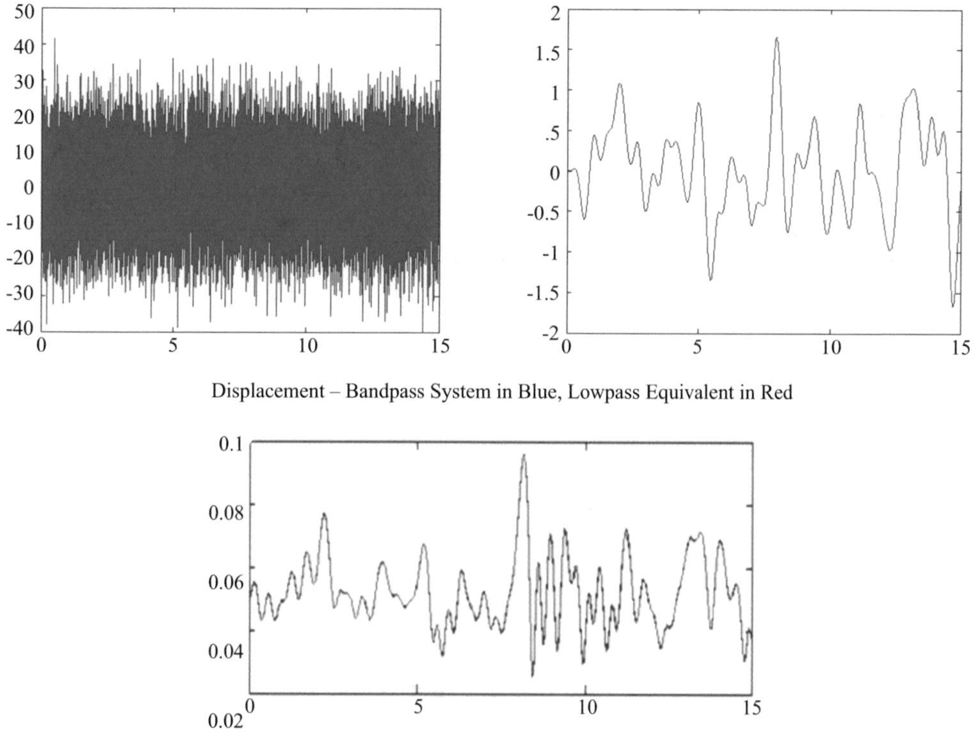

Figure 4.2: System response in the time domain (horizontal axes in seconds) to filtered (top-right) white noise (top-left) disturbance excitation d in terms of mass displacement in m (bottom); BP model response is in blue, LP in red.

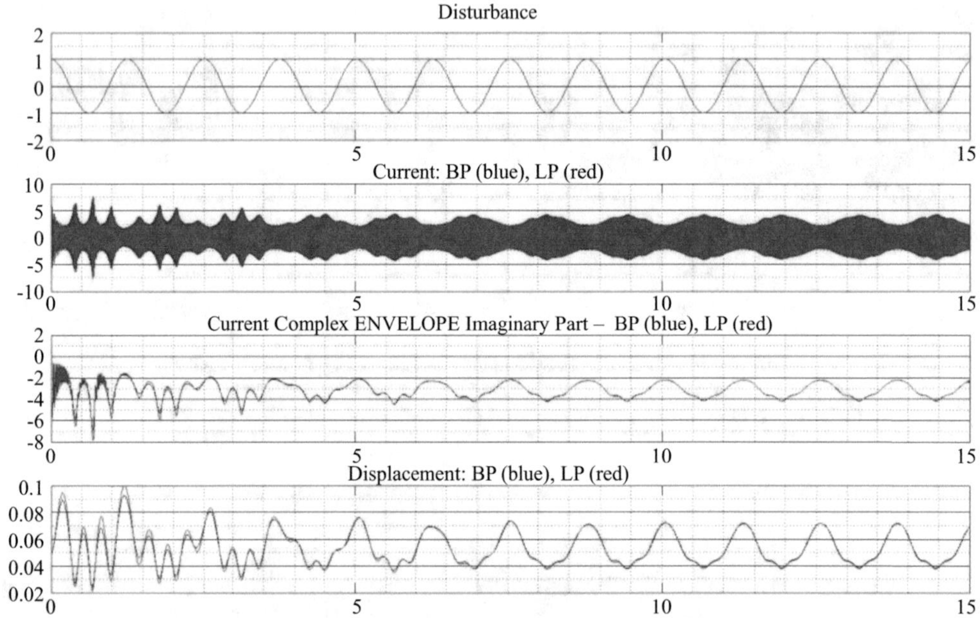

Figure 4.3: System response in the time domain (horizontal axes in seconds) to disturbance excitation $d_{\{1\}}$; the first plot is the disturbance in N, the second is the current obtained by the BP model in A, the third is the imaginary part of the complex envelope as obtained by the LP model in A, and the fourth is the mass displacement in m; BP model responses are in blue, LP in red.

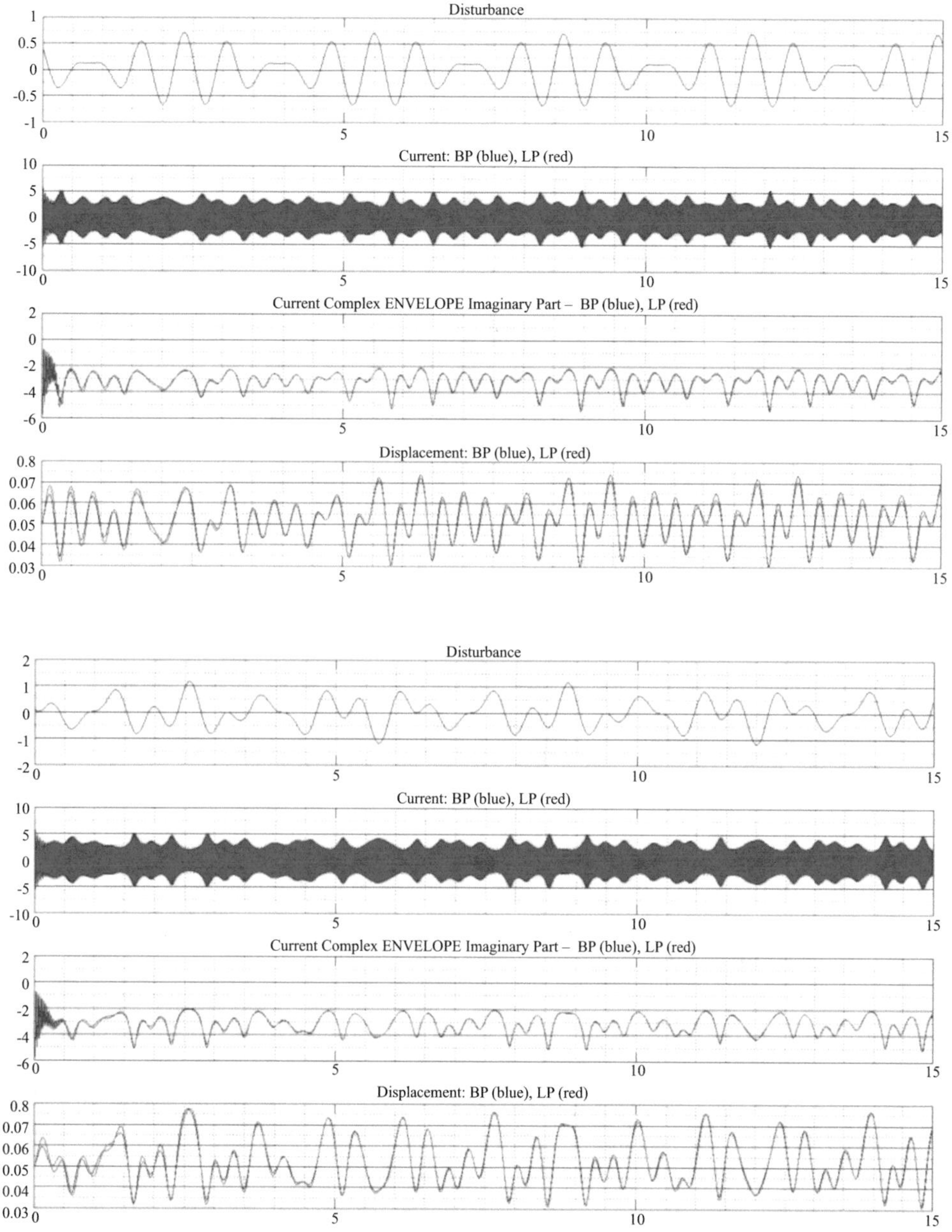

Figure 4.4: System response in the time domain (horizontal axes in seconds) to disturbance excitation $d_{\{2\}}$ (top) and $d_{\{3\}}$ (bottom); the first plot in each section is the disturbance in N, the second is the current obtained by the BP model in A, the third is the imaginary part of the complex envelope as obtained by the LP model in A, and the fourth is the mass displacement in m; BP model responses are in blue, LP in red.

Table 4.2: Test Case II settings

m (kgr)	0.25
b (kgr/s)	$25 \cdot 10^{-4}$
k (kgr/s^2)	$25 \cdot 10^{-4}$
ω_M (rad/s)	0.1000
L_0 (H)	$5 \cdot 10^{-2}$
L_1 (H/m)	$5 \cdot 10^{-1}$
C (F)	10^{-6}
R (Ohm)	10^{-4}
ω_E (rad/s)	$4.4721 \cdot 10^{3}$
$2A$ (V)	0.200, 0.400, 0.600
Ω (rad/s)	from 0 to 5 with step 10^{-3}

envelope of the source voltage maintained fixed, the device can be used as mass position sensor. This is one of the various possible operating regimes for the device. In a future work, the framework using the Volterra–Wiener theory of nonlinear operators [29, 30] for the implementation of various operating regimes of practical interest will be developed further.

4.5 TEST CASE II

The methodology for the modulating envelope dynamics of the electromechanical system under consideration was investigated further by means of a second typical example. The settings for the example are given in Table 4.2.

A numerical simulation model has been set up in order to obtain results directly in the time domain, by use of approximate integration in the Matlab/Simulink© environment. The simulation model is shown in Fig. 4.5, while typical time series for the current, obtained for different frequencies of the modulating envelope of the excitation voltage, are given in Fig. 4.6.

As can be seen in the results of Fig. 4.5, the bandwidth of the system is limited well below 2 rad/s. Indeed, a modulating envelope for the system current can be traced only for low-frequency modulating envelopes of the excitation voltage. As can be seen, the LP equivalent is sufficiently producing the essential system dynamics and therefore it can be used for further investigations.

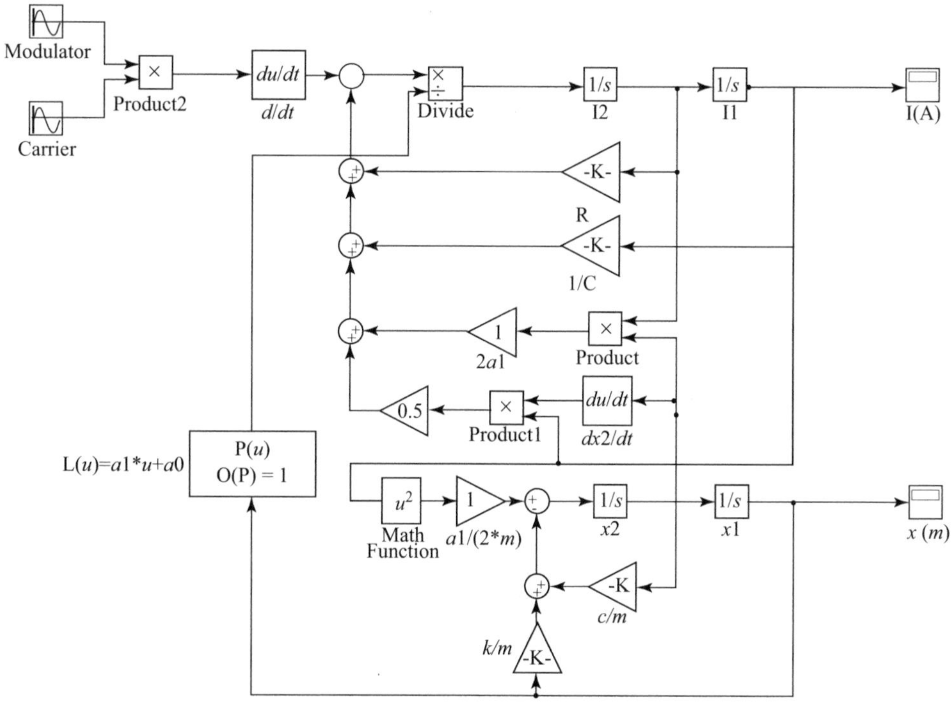

Figure 4.5: Numerical simulation model, developed in in Matlab/Simulink©, for the electromechanical system of Test Case II.

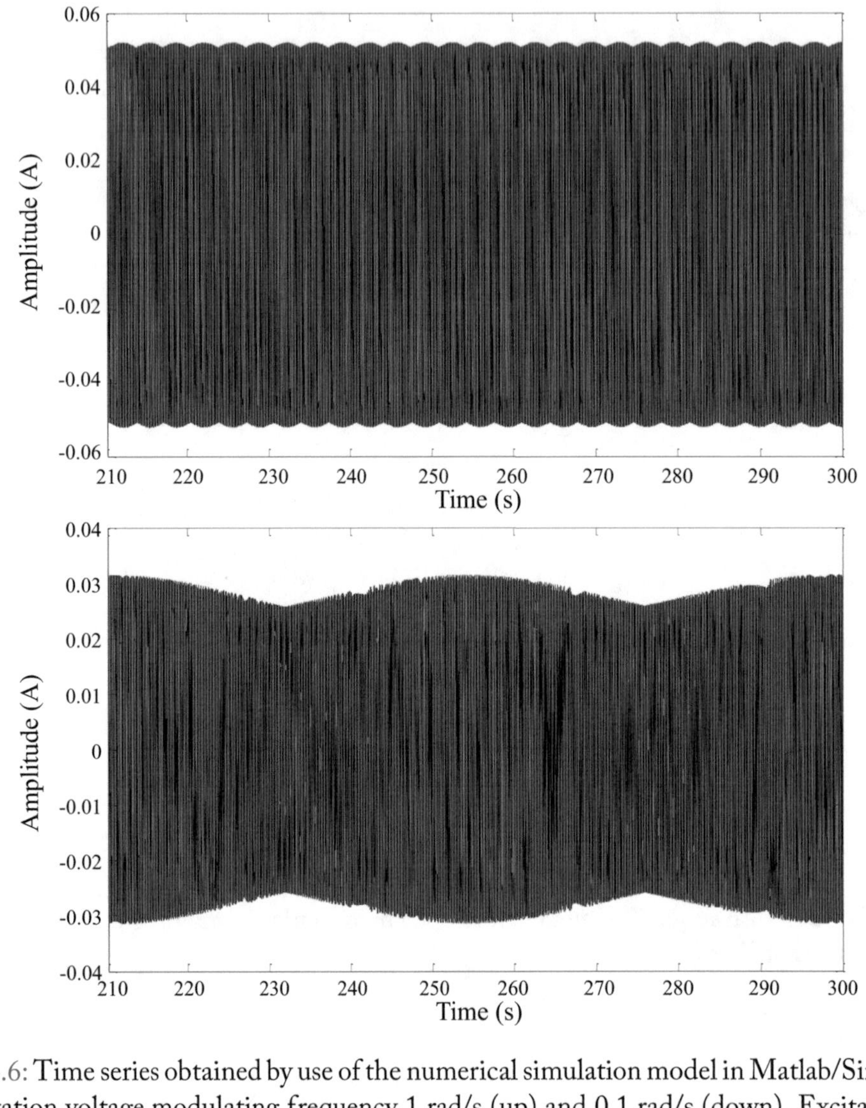

Figure 4.6: Time series obtained by use of the numerical simulation model in Matlab/Simulink© for excitation voltage modulating frequency 1 rad/s (up) and 0.1 rad/s (down). Excitation amplitude was 0.4 V in both cases.

CHAPTER 5

Conclusions and Future Work

5.1 MAIN RESULTS, DISCUSSION, AND CONCLUSIONS

This work investigated some aspects of nonlinear systems with state modulation and demodulation. As seen, even if such system consists of two spectrally decoupled subsystems for which no way exists to interact through linear dynamics, state modulation and demodulation may enable nonlinear coupling.

For this class of systems, in Chapter 1 an appropriate Hilbert Transform framework was established allowing using instead of the original equations those of the LP equivalent. The latter can be solved using a much coarser grid of time instants, if numerical methods are employed.

In Chapter 2, the general method of exact feedback linearization for input-state and input-output applications is presented. The limitations of the methodology are outlined and the relationship with the structure of nonlinear systems exhibiting state modulation and demodulation is demonstrated.

In Chapter 3, an appropriately modified exact linearization approach is derived. This is applicable to the BP subsystem of state-modulated systems with a LP and a BP subsystem. As explained, the technique is better suited for the exact linearization of this class of systems because they exhibit coupling between spectrally separated parts through space modulation and demodulation.

Finally, in Chapter 4, the techniques developed are applied to an electromechanical system consisting of a linear mechanical oscillator coupled to a linear tuned circuit. The two subsystems are spectrally separated since the resonance frequency of the tuned circuit is much higher than the bandwidth of the mechanical oscillator; furthermore, the driving input to the overall system is a voltage source with a carrier at a high frequency modulated by an envelope in a baseband compatible to the mechanical oscillator. Exact linearization is applied in this case and the value of the LP equivalent in further analysis for design and control of the system is investigated via numerical simulation.

5.2 FOLLOW-UP RESEARCH AND FURTHER WORK

The dynamics and the structure of both the general system class as well as the specific electromechanical system dealt with in this text are of particular interest for further investigations.

In mechatronics and applications of Micro- and Nano-Electro-Mechanical Systems (MEMS/NEMS), modules with the general structure like the one presented here arise rather often. The capability of using different carrier frequencies to drive on the same line and possibly by the same source a series of transducers for sensing, actuation, and control is critical for advanced instrumentation applications spanning diverse fields like maritime, automotive, aerospace, etc.

Practical considerations including determining what is the optimal (parsimonious) spectral separation between different units, the value of the carrier frequencies as well as the spatial configuration of transducer elements, should be investigated further incorporating the findings of the present work.

On a theoretical level, investigation and establishment of the conditions that can produce the structure of the BP subsystem allowing for the procedure of exact linearization developed to be applied would be a great addition to this work. Furthermore, the relationship of BP subsystem exact linearization to standard input-output and input-state linearization of the entire system is a subject deserving further research and analysis. Meaningful ways to extend the exact linearization to incorporate the LP subsystem is also very relevant and could be based on the work presented here. Specifically, Carleman linearization [29, 30] along with other techniques can be considered in this end.

At a middle level between theory and practice, questions like how to build state observers and estimation filters for the state feedback needed in exact linearization are rather worthy of further investigation. Finally, the limitations and benefits of using a synthetic input, especially if the design is based on the LP equivalent, to shape outer loops using standard linear control theory is of great interest and value.

Bibliography

[1] Xiros, N. I. and Georgiou, I. T. (2005). Analysis of coupled electromechanical oscillators by a band-pass, reduced complexity, volterra method, *Proc. of ASME-IMECE*, November 5–11, Orlando, FL. DOI: 10.1115/imece2005-81487. 1, 2, 43

[2] Xiros, N. I. and Dhanak, M. R. (2009). Control synthesis of a nonlinearly coupled electromechanical system by a reduced complexity, volterra method, *ASNE Intelligent Ships Symposium*, VIII, May 20–21, Philadelphia, PA. 2

[3] Landa, P. S. (1996). *Nonlinear Oscillations and Waves in Dynamical Systems*, Kluwer, The Netherlands. DOI: 10.1007/978-94-015-8763-1. 2

[4] Barth, H. (2003). *Sensors and Sensing in Biology and Engineering*, 2nd ed., Springer-Verlag, Austria. DOI: 10.1007/978-3-7091-6025-1. 41

[5] Rhoads, J., Shaw, S. W., and Turner K. L. (2010). Nonlinear dynamics and its applications in micro- and nanoresonators, *Journal of Dynamic Systems, Measurement, and Control*, 132:1–14. DOI: 10.1115/dscc2008-2406. 41

[6] Younis, M. I. (2011). *MEMS Linear and Nonlinear Statics and Dynamics*, Springer, New York. DOI: 10.1007/978-1-4419-6020-7. 1, 41

[7] Krause, P. C., Wasynczuk, O., and Sudhoff, S. D. (2002). *Analysis of Electric Machinery and Drive Systems*, 2nd ed., Wiley Inter-Science. DOI: 10.1109/9780470544167. 2

[8] Tu, Loring W. (2010). *An Introduction to Manifolds*, 2nd ed., New York, Springer. DOI: 10.1007/978-1-4419-7400-6. 5, 20, 21

[9] Khalil, H. K. (2000). *Nonlinear Systems*, 3rd ed., Prentice Hall. 13, 20, 21, 22, 27, 28, 32, 44

[10] Golnaraghi, F. and Kuo, B. C. (2010). *Automatic Control Systems*, John Wiley & Sons. 23, 24, 26, 41

[11] Qiu, L. and Zhou, K. (2009). *Introduction to Feedback Control*, Prentice Hall. 13

[12] Ogata, K. (1997). *System Dynamics*, 3rd ed., Prentice Hall.

[13] Bequette, B. W. (2003). *Process Control; Modeling, Design and Simulation,* Prentice Hall.

[14] Xiros, N. I. (2002). *Robust Control of Diesel Ship Propulsion,* Springer. DOI: 10.1007/978-1-4471-0191-8. 46

[15] Dhanak, M. R. and Xiros, N. I. (Eds.) (2016). *Springer Handbook of Ocean Engineering,* Springer-Verlag. DOI: 10.1007/978-3-319-16649-0. 46

[16] Giurgiutiu, V. and Lyshevski, S. E. (2009). *Micromechatronics; Modeling, Analysis, and Design with Matlab*, 2nd ed., CRC Press. DOI: 10.1201/b15830. 23, 24, 26

[17] Isidori, A. (1995). *Nonlinear Control Systems*, 3rd ed., Springer (Communications and Control Engineering). DOI: 10.1007/978-3-662-02581-9. 24, 26, 27, 28

[18] Nijmeijer, H. and van der Schaft, A. (1990). *Nonlinear Dynamical Control Systems,* Springer-Verlag, New York. DOI: 10.1007/978-1-4757-2101-0.

[19] Sontag, E. D. (1990). *Mathematical Control Theory; Deterministic Finite Dimensional Systems*, 1st ed., Springer-Verlag, New York (Texts in Applied Mathematics). DOI: 10.1007/978-1-4612-0577-7. 24, 26, 27, 28, 32, 44

[20] Xiros, N. I. and Georgiou, I. T. (2015). A Low-pass-equivalent, state-space model for the nonlinear coupling dynamics in mechatronic transducers, *Journal of Applied Nonlinear Dynamics (JAND)*, 4(1):21–42. DOI: 10.5890/jand.2015.03.003. 37, 38, 43

[21] Xiros, N. I. (2015). Nonlinear dynamic analysis for control of electromechanical systems with coupled oscillators, *Journal of Mechatronics*, 3:1–16. DOI: 10.1166/jom.2015.1092. 43

[22] Rylander, T., Ingelström, P., and Bondeson, A. (2013). *Computational Electromagnetics*, 2nd ed., Springer-Verlag, New York (Texts in Applied Mathematics, 51). DOI: 10.1007/978-1-4614-5351-2.

[23] Griffiths, D. J. (2017). *Introduction to Electrodynamics*, 4th ed., Cambridge University Press. DOI: 10.1017/9781108333511.

[24] Jackson, J. D. (1999). *Classical Electrodynamics*, 3rd ed., New York, John Wiley & Sons. 37, 38

[25] Haykin, S. (2000). *Communication Systems*, 4th ed., Wiley. 46, 47, 50

[26] Shanmugam, K. S. (1979). *Digital and Analog Communication Systems*, John Wiley & Sons. 46, 47

[27] Proakis, J. G. and Manolakis, D. G. (1996). *Digital Signal Processing; Principles, Algorithms and Applications*, 3rd ed., Prentice Hall.

[28] Bose, N. K. (1985). *Digital Filters; Theory and Applications*, North-Holland. 50

[29] Rugh, W. J. (1981). *Nonlinear System Theory; The Volterra/Wiener Approach*, The John Hopkins University Press. 54, 58

[30] Schetzen, M. (1980). *The Volterra and Wiener Theories of Nonlinear Systems*, John Wiley & Sons. 54, 58

Author's Biography

NIKOLAOS I. XIROS

Nikolaos I. Xiros was born in Athens, Greece. His career spans more than 20 years in both industry and academia, and his expertise lies within the fields of marine and electromechanical systems engineering. He holds an Electrical Engineer's degree, an M.Sc. in Math, an M.Sc. in Physics, and a Marine Engineering Doctorate. His research interests are process modeling and simulation, system dynamics, identification and control, reliability, and signal and data analysis. He has authored many technical papers and the monograph *Robust Control of Diesel Ship Propulsion*. He is also Chief Editor of the *Springer Handbook of Ocean Engineering*.

Printed in the United States
by Baker & Taylor Publisher Services